JET
THE ENGINE THAT CHANGED THE WORLD

GRAHAM HOYLAND

Books

Published by Key Books
An imprint of Key Publishing Ltd
PO Box 100
Stamford
Lincs PE9 1XQ

www.keypublishing.com

Originally published in 2022.
This paperback edition was first published in 2023.

ISBN 978 1 80282 678 4

Front cover image: Thom Lang/Getty Images
Cover design: Myriam Bell Designs
Back cover image: Key Archives

Typeset by SJmagic DESIGN SERVICES, India.

Acknowledgements

I am writing these words aboard an Airbus A320. Thank you to Frank Whittle who made flight safe, fast, comfortable and affordable. And once again, I have to thank my friend and neighbour, Bryan McGee, a member of the Rolls-Royce Owners' Club, who generously loaned me most of the books listed in the bibliography, and who provides endless support to my writing endeavours. Above all, I would like to thank my partner Gina Waggott, who gives me endless encouragement and support.

Contents

Take Off

The invention of the jet engine had a profound social effect on the whole world. Commercial jet aircraft revolutionised travel, opening up every corner of the planet. Millions were able to travel anywhere in the world for the price of a month's salary.

The dream of cheap exotic travel might have been realised, but there would be far-reaching consequences of shrinking the world with the new technology. This book tells the story of jet technology, how it enabled mass transport by air and how that changed the world we live in.

Jet engines came on the scene just when conventional piston engines had reached their physical limits, being made from thousands of jerking, reciprocating and overheating components. In a masterstroke of simplification, the jet engine replaced all of these with one moving part. The piston engine's processes of sucking, squeezing, banging and blowing took place continually in different parts of the jet, instead of sequentially in one place. The result was a smooth flow of power.

By 2020, people of modest means, such as a Filipina nurse working in Kuwait or a Nepali construction worker in Saudi Arabia, could travel home to see their families. But the speed of human travel around the world also encouraged the spread of a lethal pandemic. Flights stopped overnight, and dozens of airlines went bust. The Boeing 747 and the Airbus A340 were retired early. Airline easyJet warned that aviation faced the most severe threat in its history.

Like a runaway jet engine, humanity's appetite for travel had grown until it had exploded. This may well be the end of the era of cheap flights. It's time to appraise the Age of the Jet.

Chapter 1

The Attainment of Power

The history of mankind is the history of the attainment of external power. Man is the tool-using, fire-making animal.[1]

Frank Whittle's jet engine ran for the first time on 12 April 1937. It turned out to be rather more exciting than he expected.

For a second or two the speed of the engine increased slowly and then, with a rising shriek like an air-raid siren, the speed began to rise rapidly, and large patches of red heat became visible on the combustion chamber casing. The engine was obviously out of control… I screwed down the control valve immediately, but this had no effect and the speed continued to rise…[2]

Surely never in the history of engineering had such a simple invention replaced such a complicated one. By 1929, the piston aero engine had become a mechanical monstrosity: the Rolls-Royce Schneider Trophy-winning R engine of that year was made of around 14,000 separate parts, most of them moving and all of them vital to the working of the whole.

---

1 Wells, H. G., *The World Set Free*, Macmillan, London (1914)

2 Whittle, Frank, *Jet*, Muller, London (1953)

In that same year, a 21-year-old British RAF cadet dismissed all this complication at a stroke by replacing the thousands of moving parts with just one: a turbine. It was a masterpiece of upside-down thinking: suddenly the voids between pieces of metal became more important than the pieces themselves. It wasn't so much 'less is more' as 'nothing is more'. The new jet engine would eventually produce a hundred times more power than the old piston engine, and, on the way, it would completely change the world.

With the benefit of hindsight, it is clear to see that Frank Whittle was a genius. His name should stand with those of Tesla, Curie and Franklin. Not only did he conceive of his invention, but he also doggedly drove it through to fruition against the most extraordinary barrage of bureaucratic resistance, intellectual theft and commercial greed. If he had been given support in 1930, as he so nearly was, his jet engine might even have prevented the Second World War. In a conversation with Whittle after the war, Hans von Ohain, his German rival jet pioneer, said, 'If you had been given the money you would have been six years ahead of us. If Hitler or Göring had heard that there is a man in England who flies 500mph in a small experimental plane and that it is coming into development, it is likely that World War II would not have happened.'

Few know that it was Whittle's engine that introduced America to the Jet Age, or that, despite losing seven years in failing to gain any support, he still gave Britain a three-year lead in jet technology after the Second World War.

Since then, the jet engine has changed the world in the most extraordinary fashion, and, in so doing, Whittle realised for millions of us the eternal dream of humanity: the surreal experience of flight.

But in spite of his many talents, Frank Whittle had one crippling deficiency which blinded him to the worse aspects of people: naivety.

Any child shown an inflated balloon with the neck pinched between finger and thumb knows what fun it is when the pressurised air inside is allowed to escape. They are also being shown a demonstration of Newton's Third Law of Motion: as the air escapes from the balloon, it exerts a force on the air outside the balloon. As a result, the balloon is propelled forwards by the opposing force. This force is thrust, and thrust is what jet engines are all about.

A child's toy – a steam engine – inspired the young Albert Einstein to follow a life of science. A toy aeroplane given to the Wright brothers led to their invention of powered flight. And it was also a toy aeroplane given to the four-year-old Frank Whittle that led to his invention of the jet engine. 'Its propeller was driven by clockwork and it was much too heavy for a flying model, but it could be suspended from the bracket of the gas mantle by a length of string and it would perform a circle in the brief interval before the spring ran down…'[3]

Frank Whittle was born in 1907, at just about the same time as the Honourable Charles Rolls was becoming interested in aeroplanes. Having founded a motor car company with Henry Royce that was now making 'the best car in the world', the 40/50 Silver Ghost, Rolls tried to persuade his partner to make piston engines for aircraft. Royce was reluctant, having no idea that their company would go on to become the foremost manufacturer of aircraft engines in the

---

3 Whittle, Frank, *Jet*, Muller, London (1953)

world. All that was far in the future, and Rolls was to die in just three years' time, the first Briton to be killed in an aeronautical accident with a powered aircraft.

Like Henry Royce, Frank Whittle was born into an English working-class family and became a world-class engineer. The Whittles had left the cotton mills of Lancashire and moved to Coventry, where Frank's father, Moses, worked as a foreman in a factory making machine tools. He was a prolific inventor, whose many ideas were held back by inadequate schooling: he had started working in the mills at the age of ten. He was a committed Wesleyan. He also beat the unruly Frank with a razor strop when necessary.

Frank, therefore, grew up in a disciplined but inventive environment and became fascinated by aircraft, particularly when one force-landed on open ground near to his home. It was the first time he'd been able to inspect a flying machine at close quarters. He got so close it nearly removed his head when it took off again; his cap had to be retrieved from a gorse bush.

His father, Moses, eventually scraped the money together to start a small business: the Leamington Valve and Piston Ring Company, where young Frank started undertaking piecework such as drilling valve stems and lathe operations. From the beginning, he was immersed in the business of shaping engine parts, something that would stand him in good stead when he was making components no one had attempted to make before.

Then, during the First World War, something happened that had a formative effect on Frank. The British B.E.2 biplane was a design by Geoffrey de Havilland, the chief designer and test pilot at Royal Aircraft Factory at Farnborough. The unit cost was £2,061, or around £240,000 today. After a successful beginning to its operational life, shooting down Zeppelin airships, by late 1915, this aircraft had fallen behind the more powerful and manoeuvrable German fighters. The B.E.2's engine was a V-8 air-cooled Renault design built under licence, amongst others by Rolls-Royce (who would have imagined Rolls-Royce building engines for Renault?). Although the capacity was 8.8 litres, the power was an underwhelming 92hp. This was not enough to get away from pursuing German aircraft, and by mid-1916, no less than 120 British aircraft had been shot down during the Fokker Scourge. The British public was furious and frightened. The importance of having more engine power than your enemy was firmly impressed on the young boy.

At the end of the First World War, the Whittle business struggled with a lack of orders, and, for a while, the family was homeless and haunted by poverty. They had to sleep in the workshop, which was just a room with a couple of lathes. Frank was 'a street urchin on six days a week and a carefully dressed little boy on Sunday',[4] but he started to do well at school, where he dreamed of being a pilot. He won a scholarship to Leamington College, and he decided he wanted to join the Royal Air Force as a boy apprentice.

He entered and passed the written examination and thought the medical would be a formality. However, he was rejected for having a poor physique, being under-sized at 5ft (152cm) tall with a correspondingly small chest measurement. Bitterly disappointed, he begged for reconsideration:

> My persistence, however, was not entirely wasted. A service policeman took me along to see a physical training instructor, Sergeant Holmes, who also took pity on me. He gave me a list of Maxalding exercises and showed me how to do them. In addition, he wrote out a diet

---

4 Whittle, Frank, *Jet*, Muller, London (1953)

sheet for me, and when I returned to Leamington, I carried out his instructions to the letter.[5]

Fortified with daily doses of cod liver oil and bulging with newly acquired muscles, Whittle added three inches to his height and three to his chest measurement. Returning for a second medical examination, he was turned down again. Undaunted, he hit on the ruse of applying again as if he had never applied before and, once again, passed the written examination. Then, at last, he passed the medical. He was finally in the RAF, a boy apprentice at the age of 16.

Lord Trenchard had set up the Apprentices School as part of a development for the new RAF. It was to supply a flow of skilled technicians, without whom it could not operate. After three years of grinding discipline and a first-class training as a fitter, 364365 Apprentice Whittle, F. was one of only five boy entrants out of 600 to win a cadetship to the RAF College at Cranwell, and his dream of becoming a pilot became feasible. It was at this point that social class reared its ugly head.

'It is impossible for an Englishman to open his mouth without making some other Englishman hate or despise him.' So wrote the Irish outsider George Bernard Shaw in his preface to *Pygmalion*, his 1912 play on English social climbing. With a plot lifted straight out of Smollett's *Peregrine Pickle*, Shaw's satirical commentary on English class and diction remains his most popular work, and it became a musical and a film: *My Fair Lady*. Professor Higgins is a linguist, able to deduce a person's class and circumstances from hearing them speak. He overhears Eliza Doolittle, a common London flower girl, and wagers he can train her to pass as a duchess. The rest is predictable, except for the absence of a happy ending. To his credit, Shaw preferred Eliza's new self-determination to a conventional

---

5 Whittle, Frank, *Jet*, Muller, London (1953)

outcome. One actor playing Higgins sought to sweeten Shaw's ending by tossing a bouquet to Eliza, implying marriage. Shaw was furious. 'My ending makes money; you ought to be grateful,' protested the actor. To which Shaw replied, 'Your ending is damnable; you ought to be shot.'[6]

Like Eliza Doolittle, Frank Whittle yearned to lift himself out of his social class, and at Leamington College he went to some trouble to rid himself of his Lancashire accent. He would sit near Leamington High School girls at play, listening carefully and imitating their mode of speech. But he was physically small, he disliked team games, and he didn't fit in with the self-confident and team-spirited ex-public school boys of Cranwell. His family had to scrape up the money to kit him out with a dark lounge suit and bowler hat, and he and his first-term colleagues would march together, heads up, arms swinging, with battered bowler hats on their heads, many with the crowns detached and flopping up and down in time with the marching step. Although his apprenticeship training gave him advantages in the workshops, the social gap was wide. The other boys regarded him as 'common'.

One night, Whittle was lying in bed after lights out and heard approaching footsteps; he was told to report to Hut 22:

> On arrival I found the whole of the Third Term sitting in rows facing a table. 'Get on that table,' I was ordered, and then standing to attention in pyjamas, I learned that I had been walking around as if I owned the place and that it was time I had the bounce knocked out of me. I was then ordered to leap off the table and bend over a chair to receive one swipe from every member of the court with whatever implement

---

6 *The Guardian* (10 Feb 2004)

took their fancy. It was most important not to make a sound during this procedure, though this was extremely difficult when some of the more sadistic members took their turn. I received some twenty blows – though quite a few of them from some of the softer-hearted members of court were comparatively gentle.

Nobody was more surprised than I was to learn that I had been walking about as though I owned the place, because I supposed I had been creeping about as inconspicuously as possible, not only acutely conscious of the fact that I was a First Termer but also one of the few ex-apprentices.[7]

Despite the bullying, RAF Cranwell was where Whittle's far-sighted genius started to blossom. He became engrossed in mathematics, physics and the theory of flight. Then, in 1928, at the age of only 21, he wrote a thesis titled 'Future Developments in Aircraft Design'. In it, he forecast speeds of over 500mph (800km/h), flying above 33,000ft in the stratosphere. At the time, the latest frontline fighter planes seen at Cranwell were piston-engined biplanes with a maximum speed of less than 150mph (240km/h) and a service ceiling of only 20,000ft.

Whittle had studied the properties of the atmosphere in an earlier thesis and learned something important. The stratosphere is the second layer of the Earth's atmosphere, lying above the troposphere and below the mesosphere. The stratosphere is stratified, or layered, in temperature, with the warmer layers higher and the cooler layers closer to the Earth; this increase of temperature with altitude is a result of the absorption of the sun's ultraviolet radiation by the ozone layer. This is in contrast to the troposphere, next to the Earth's surface where, as any mountaineer knows, temperature decreases

---

7 Whittle, Frank, *Jet*, Muller, London (1953)

with altitude. Near the equator, the lower edge of the stratosphere is as high as 66,000ft, around 33,000ft at midlatitudes, and at about 23,000ft at the poles. These are the heights at which modern jet airliners fly. And they fly there for a good reason.

Whittle knew that at those altitudes the air density was one-quarter of its sea level value, and so the air's resistance to an aeroplane was correspondingly diminished. The potential speed and range could thus be far higher. If an engine could be persuaded to run in these low air densities, then aircraft could have a crushing advantage over their enemies below: aerial dogfights during the First World War had shown, above all else, that superior speed and altitude were a winning combination.

> I came to the general conclusion that if very high speeds were to be combined with long range, it would be necessary to fly at very great height, where the low air density would greatly reduce resistance in proportion to speed… It seemed to me unlikely that the conventional piston engine and propeller combination would meet the power plant needs of the kind of high-speed/high-altitude aircraft I had in mind, and so in my discussion of power plant I cast my net very wide and discussed the properties of rocket propulsion and of gas turbines driving propellers, but it did not then occur to me to use the gas turbine for jet propulsion.[8]

Whittle and a small number of aero-engineers recognised that the piston engine and propeller were providing diminishing returns. Despite reductions in drag and increases in efficiency provided by slippery airframes and variable-pitch propellers, huge increases in engine horsepower were yielding smaller and smaller increases

---

8 Whittle, Frank, *Jet*, Muller, London (1953)

in speed. And as propeller-driven aircraft approached the speed of sound, they came up against what the historian of technology Professor Edward Constant[9] calls a presumptive anomaly: progress in a given line of technological development will always be capped by a theoretical upper limit.

A propeller-driven aircraft screws its way through the air similar to the way a corkscrew twists through a cork. The tips of the propeller have to travel further and faster than the aircraft, and so when it gets close to sonic speed, the propeller begins to go supersonic, starting at the tips and working inwards. Spectators on the ground underneath Schneider Trophy aircraft would hear a deafening crashing and banging as the propeller tips exceeded the speed of sound. For the aircraft to go supersonic, the whole propeller would have to be travelling through the air much faster than the speed of sound. And there is a problem with that. The drag on the blades increases hugely as they approach Mach 1, the thrust diminishes and the noise and shockwaves become so destructive that the airframe itself is imperilled. So, there was a clear limit to the speed of piston-engined propeller-driven aircraft. This was the presumptive anomaly, and everyone in the business of aeroplanes in the 1930s knew that propeller aircraft were near the limit of what was possible.

One way you could approach the speed of sound was if your propeller fell off. And this is exactly what happened to Squadron Leader Tony Martindale in April 1944. Exploring high speeds for the RAF, he rolled over and put his Merlin-engined Mark XI Supermarine Spitfire into a dive from 40,000ft. With the propeller set to fully coarse, the engine's revolutions per minute (rpm) went off the clock, and at something approaching 600mph (970km/h), the Merlin's reduction gearbox ripped off, taking the propeller with it.

---

9 Constant, Edward, *The Origins of the Turbojet Revolution*, The Johns Hopkins University Press, Baltimore (1980). Professor Constant was famously combative, shooting a police officer after a drunken domestic argument. After himself being shot in the buttocks by the enraged officer, Constant was arrested and sentenced to 29 years in prison.

The Spitfire then reached over 620mph (1,000km/h) – Mach 0.92 – as it plunged towards the ground. With the heavy gearbox and propeller missing, Martindale's Spitfire was now tail-heavy and so this change in the centre of gravity forced it into a steep climb. Martindale briefly lost consciousness due to positive G and woke up to find his aircraft flying quietly along on its own, without a propeller.

It could be argued that this Spitfire had just become a jet aeroplane, as it had a Meredith-effect radiator providing jet thrust, and the Merlin engine had backwards-pointing exhaust stubs that did the same.

A consummate pilot, Martindale glided the Spitfire back to base and got out somewhat shaken. His ground crew pointed out there was even more damage: the speed of the Spitfire's dive had slightly bent the wings backwards. This was amazing. Mach 0.92 was far higher than anything reached by any other piston-engined aircraft.[10]

Whittle started to speculate on the kind of engine that could operate at stratospheric altitudes. Up until the end of the Second World War, the predominant power plant for aircraft was the Otto-cycle internal combustion piston engine.

To those of you who have read my book about the Rolls-Royce Merlin engine,[11] I apologise for the following, but it might be helpful to remind ourselves what a revolution the jet engine was. The way an Otto-cycle four-stroke petrol engine works is the same in a piston-driven aircraft or a car engine: a piston, which looks like a soup can with one end removed, slides up and down a tube or cylinder. The piston is connected to a rod called, unsurprisingly, a

10 See Gunston, Bill, *Faster Than Sound,* Patrick Stephens, Wellingborough (1992) for a fuller discussion of the sound barrier.

11 Hoyland, Graham, *Merlin: The Power Behind the Spitfire, Mosquito and Lancaster,* William Collins, London (2020)

connecting rod. This is attached to a crankshaft so that the piston sliding up and down pushes the crankshaft around as a cyclist's leg pushes a pedal crank around. The crankshaft can be connected to car wheels or an airscrew propeller.

Above the piston is where the magic of internal combustion takes place. The top of the cylinder is closed but is provided with an inlet valve to let in a mixture of air and fuel as the piston descends on the first, or inlet stroke. This valve then closes when the piston reaches the bottom and starts up again on the second, or compression, stroke. The gaseous mixture of air and fuel is now squeezed tightly and is highly explosive. At the top of the compression stroke, a sparking plug ignites the mixture, and it duly explodes. The piston is now pushed down on its third stroke, the power stroke. On its way back up, an exhaust valve opens, and the hot gas is forced out by the ascending piston on the fourth, exhaust, stroke. A glance at an online animated graphic will explain it all. Just think, 'suck, squeeze, bang, blow'.

Otto's inventive step was to manage these four distinct phases inside a cylinder with a piston travelling up and down, and he did this by arranging the valves to open at specific times during two revolutions of the crankshaft, or four strokes of the piston. This was done by using a camshaft, which looks something like a knobbly stick, to push the valves open with oval-shaped lobes. By making the camshaft rotate only once for every two rotations of the crankshaft, the valves open at the correct times. This whole mass of machinery spins round at up to 100 times a second.

The disadvantages of pistons engines are many. Firstly, the complication of thousands of parts adds weight, expense and unreliability. Secondly, making all four cycles happen sequentially in the same small chamber leads to vibration, overheating and a restricted throughput of fuel and air. Fuel and air are fighting to get in and burnt exhaust gases are fighting to get out. Thirdly, there is only one power stroke; the other three strokes merely serve to wear the engine out. Fourthly, the reduced air density at altitude drastically reduces the engine power, as the numbers of oxygen molecules necessary for combustion are reduced in proportion.

Whittle knew that for a piston engine to make significant power at high altitudes, it would have to be supercharged. Mountaineers carry bottled oxygen to the summit of Mount Everest because the air density there is a third of that of sea level, and the oxygen levels are correspondingly low. Aircraft, though, need vast amounts of oxygen: the contemporary Rolls-Royce R piston engine would ingest the volume of a single-decker bus in just one minute. And a modern jet engine sucks in as much as two tons of air per second during take-off; that's about the volume of a swimming pool. Aircraft cannot carry enough bottled oxygen, but they *can* force extra air into a piston engine using an air pump called a centrifugal supercharger, or blower. This is a disc with vanes on it that can be driven directly from the crankshaft (or by a turbine in the stream of exhaust gases, in which case it is called a turbocharger). This supercharger disc, or impeller, was going to be a crucial part of Whittle's future design.

Whittle's hand-written thesis now looks school-boyish, with pages of densely scrawled formulas. In his summary, he proposes a new form of prime mover, or power plant, consisting of a piston engine combined with an air compressor feeding hot gases into a pressure tank, which, in turn, revolves turbine blades, which then turn a propeller.

'It seems that, as the turbine is the most efficient prime mover known, it possible that it will be developed for aircraft especially if some means of driving a turbine by petrol could be devised. A steam turbine is quite impractical owing to the weight of boilers, condensers etc.'[12]

---

12 Whittle, Frank, *Jet*, Muller, London (1953)

His first proposal would also have been far too heavy for flight, as it added an air compressor and a thick-walled pressure tank to an existing piston engine. This is about as far as Whittle got with his ideas in his thesis. He was awarded full marks by his chemistry professor, O. S. Sinnatt, who said, 'I couldn't quite follow everything you have written, Whittle. But I can't find anything wrong with it.' His thesis was to change the face of modern civilisation.

Whittle's main ambition was to become an outstanding pilot, and in that he succeeded spectacularly – indeed, rather too spectacularly. He was disqualified for dangerous flying during an aerobatic show after performing the legendary Bunt: the first half of an outside loop, a steepening dive that ends with the aircraft inverted. This dangerous-looking manoeuvre demands a great deal of courage and commitment from the pilot. During his performance in front of judges, Whittle throttled back his Armstrong Whitworth Siskin fighter and pushed the stick forward. His flying speed increased as he dived for the ground. Only his harness straps, pulled up tight, kept him in the cockpit. Negative G built up and increased the blood pressure in his brain as he completed the bottom of the loop close to the ground upside down, before half-rolling into level flight, just above the grass.

At the end of his course, Whittle was assessed as 'Exceptional to Above Average' on the Siskin, but inscribed in his flight log was this comment in red ink:

Over Confidence. This is most marked. He gives aerobatics too much value and has neglected accuracy. Must learn to discipline his flying. Inclined to perform to the gallery and flies too low.

On 27 August 1928, Frank Whittle joined his first fighter squadron and so became an officer and a gentleman. But he nearly wrecked his career with the RAF by continued dare-devil flying.

When these stunts are examined now, you can see he was applying his genius to the calculation of risk. Some of us are born with their risk thermostat set higher than others, but Frank Whittle's risk thermostat was jammed at around 11.

Pilots in Fighter Command could largely please themselves with what they did with their flying time, and Whittle delighted in low-flying stunts along the Thames and pushing the performance envelope of the Siskin III biplanes. He had been making careful calculations of the fuel consumption of the aircraft and decided to test the endurance. When, one day, he did not return, it was assumed he had crashed, and on his return, he was reprimanded for staying in the air for longer than was possible!

Whittle also practised tail slides, where the pilot stalls the plane with the nose up, and then falls backwards towards the ground under gravity. His flight commander, Flight Lieutenant Davidson, was furious when Whittle told him the stresses on the airframe during this manoeuvre were negligible. (Davidson had a point, as any sailor knows that a backwards flow on a rudder can wrench it from the fastenings if it snatches to one side. But Whittle knew to hold the controls firmly.)

The urge to perform stunts is part of any fighter pilot's character, but Whittle had this to a marked degree. His speciality was horrifying to watch. He would perform a flick roll at ground level, rolling the whole aircraft upside down. 'I had developed it to the point where I could just spin my wheels by touching the ground immediately before doing it. It must have been a shattering sight for anybody witnessing it…'[13]

---

13 Whittle, Frank, *Jet*, Muller, London (1953)

These stunts were acceptable during formation displays but the Air Ministry took a dim view of any pilot doing them where the public might be concerned. It had become a court martial offence, punishable by dismissal from the service. The RAF was fighting for funds in a climate of disarmament, and it could not afford bad publicity created by irresponsible pilots. So, when a policeman reported Whittle and another young pilot performing low aerobatics over the Thames at Canvey Island, he thought his career was over. He visited the police station and was told a local family had complained, and that if they dropped it, the police would not press the matter. Whittle visited the Levy family, who were horrified that they could be career-wreckers. They had become concerned for their pregnant daughter's state of mind after watching the aerobatics. They agreed to drop the matter.

Unfortunately, Whittle's new commanding officer, Squadron Leader Soden, landed at Canvey Island shortly afterwards in one of 111 Squadron's planes and was immediately accosted by the same policeman, who demanded to know what the hell he thought he was doing and that they were sick of low-flying aeroplanes over Canvey Island. The cat was out of the bag. Whittle and his accomplices were ordered to present themselves before their new CO, and he picked up the police report:

> Soden commented, 'You lucky young devils!' meaning it was lucky for us the report had been sent to him and not the Air Ministry, and dropped the report in the wastepaper basket. Had he not done so the whole course of the rest of my life would have been altered.

Perhaps because of these stunts, Whittle was detailed to test catapults, which, before the advent of aircraft carriers, were used to launch aircraft off warships. Until then, the tests had been done with animals: 'The missile was the battered fuselage of an aeroplane, and strapped into the cockpit in an intrepid and professional attitude, was a large sheep – because that was the nearest approach to a Naval pilot that R.A.E. scientists could think of ... Whoosh! – The expression on the sheep's face was something the writer will never forget...'[14]

Whittle was then strapped into a real aeroplane and began a series of tests with cordite-powered catapults, which were liberally smeared with glycerine as a lubricant. He was experiencing accelerations of 2.5g. One early test went wrong when, instead of glycerine, lubricating oil was used. This exploded in the ram cylinder of the catapult and fired the unfortunate pilot 150ft into the air with an acceleration of 6g, which caused the aircraft to disintegrate around him. Somehow, he survived.

The catapult tests progressed to sea trials with real powered aircraft on a ship and on one test Whittle's passenger, Flight Lieutenant Kirk, was jerked out of the rear cockpit of the Fairey III F floatplane by the force of the catapult explosion. He slid down the fuselage on the slippery glycerine, but on the way past he just managed to grab the tailplane and hung on. As the aircraft started to fly, Whittle couldn't understand why the nose had risen sharply until he glanced back. On the brink of a stall, the floatplane staggered across the Solent at an altitude of 60ft:

> Meantime, a German liner was crossing my path ahead and the question was whether I could get down on the water before I flew into her side. I succeeded with about 200 yards to spare and alighted with the stick fully forward by easing the throttle slowly back... Kirk then clambered back into the cockpit... He seemed quite calm and collected, while I was in a state of nervous collapse. Weak and trembling with reaction, I looked round at him and just said

---

14 Bethell, Peter, Commander RN, *Engineering*, Vol. 156 (2 July 1943)

'My God!' to which he calmly responded, 'What are you worrying about?'[15]

After surviving these hair-raising stunts, Whittle was sent to flying instructor's school at RAF Wittering, where he continued his thinking around the problem of a high-altitude power plant. The next idea he had was for a motorjet, which consists of a conventional piston engine driving a compressor instead of a propeller. The compressed air and hot exhaust gases are passed into a combustion chamber where fuel is injected and ignited. The high temperatures generated by the combustion cause the gases in the chamber to expand and escape at high velocity from the exhaust, creating a thermal reactive force that provides thrust.

Then, in a blinding eureka moment, Whittle realised that the piston engine was superfluous: a turbine in the exhaust flow could drive an air compressor at the front of the engine. He literally threw away the piston engine.

And so, in October 1929 in rural Wittering, Whittle finally developed his Cranwell thesis into the answer he was searching for. His mathematics had led him back to the gas turbine, but this time it was a new type that produced a pushing jet instead of driving a propeller.

---

15 Whittle, *Jet*, p. 34. Margaret Horton was a WAAF (Fitter Mechanic Airframes) at an RAF airfield on 9 February 1945. Because the conditions were windy, Margaret was instructed to assist the pilot of Spitfire AB910 by sitting on the tailplane while he taxied to the take-off position. This was to prevent the aircraft 'ground looping'. At the end of the runway, the pilot, Neil Cox, turned into wind and, forgetting Margaret was still on the tail, started his take-off run. Margaret, realising she could not jump off safely, clung on to the Spitfire's rudder upright. The pilot took to the air but realised there was something wrong with his Spitfire. The control tower, without telling him the reason, told him to land immediately. He managed to land safely with Margaret still clinging to the tailplane. She was credited with the airtime in her logbook.

Whittle's inventive step was to substitute a compressor for a piston engine in the system, proposing a pure turbojet engine. He later expressed the need for a new engine:

> Reciprocating engines are exhausted. They have hundreds of parts jerking to and fro, and they cannot be made more powerful without becoming too complicated. The engine of the future must produce 2,000 hp with one moving part: a spinning turbine and compressor.[16]

His idea was stunning in its audacity. It involved circular thinking: the air arriving at the mouth of the engine is compressed by a device that is powered by the hot gases driving a turbine at the *end* of the process. The excess leftover exhaust flow would provide thrust to propel the aircraft along at huge speeds. No one who thought along conventional lines would have come up with this idea.

As we have seen, the low air density at high altitude reduces drag on aircraft but still provides enough ram effect to force air into the jet engine. The low air temperature at altitude also serves to increase efficiency. It was certainly a eureka moment.

It was the Greek scholar Archimedes who famously cried 'Eureka! Eureka!' ('I have found it! I have found it!') after lowering himself into a bath. He had solved a problem that had been bedevilling the city of Syracuse. So overwhelmed by his discovery, and so eager he was to share his idea, he jumped out of his bath and raced naked through the streets.

The tyrant of Syracuse, Hiero II, suspected he had been cheated by a goldsmith. He had ordered an irregularly shaped votive crown for a temple and given the man gold. Hiero guessed the goldsmith

---

16 Whittle, Frank, *Jet*, Muller, London (1953)

had mixed the gold with cheaper, lighter silver and had kept the remaining gold for himself. He asked Archimedes to find out if all the gold had been used but weighing the crown against a similar amount of pure gold proved inconclusive.

When Archimedes got into his bath, he observed that the water level rose and suddenly realised that the volume of water displaced was exactly equal to the volume of the part of his body he had submerged. The volume of irregularly shaped objects could therefore be measured with precision (and the bottom half of Archimedes was also irregularly shaped). So, two irregularly shaped items of identical volume but different density would have different weights when submerged in water.

As silver is half as dense as gold, all he had to do was once again balance the crown against pure gold on a scale in the air, and then submerge both the crown and the gold in water simultaneously. If the volumes of crown and gold were the same, the scale would remain in balance: their densities were the same and, therefore, the crown must be made of pure gold. But if the volume of the crown was greater because of the added silver, its increased buoyancy would result in imbalance. Therefore, the crown could not be pure gold.

The test was performed in public, the scales tipped up as they were lowered into the water. And then Hiero slowly turned a terrible countenance upon the cringing goldsmith.

At school, Whittle had studied Aurel Stodola's *Steam and Gas Turbines* textbook, so he was familiar with the concept of turbines.

The modern steam turbine had been invented in 1884 by the Anglo-Irishman Charles Parsons. The son of an earl, he took the unusual step of joining the Newcastle-based engineering firm of W. G. Armstrong as an apprentice. Somewhat like Whittle, he threw out the ponderous old piston steam engine and came up with the turbine, a wheel of angled fan blades mounted on a rotating shaft. A series of these wheels are fitted onto the shaft and placed inside a large tube. One set of stationary blades is connected to the casing and one set of rotating blades is connected to the shaft, interleaved rather like the fingers on clasped hands. These sets of blades intermesh with close clearances, so that when high-pressure steam is let into one end of the tube, angled jets of steam are directed onto the rotating blades and they drive them round and release enormous amounts of power. Each set of fixed and moving blades progressively exploits the expansion of the steam in stages.

In Parsons's patent application, he states that the turbine could be driven 'in reverse' to act as a compressor. He suggests using a compressor to feed air into a furnace, and a turbine to extract power to run the compressor. Although this is intended for factory use, and in no way is he suggesting pushing the furnace along, he *is* describing a gas turbine.

The spinning shaft of his steam turbine was ideal for driving generators, and Parsons's first application was for a dynamo of 7.5kW. He swiftly realised that his steam turbine could be scaled up and, therefore, would be ideal for driving ships at high speed. He tried to interest the Royal Navy in his new prime mover by building his sensationally fast steam-turbine-powered *Turbinia*, an 'experimental yacht', but he suffered from incomprehension and backwards thinking from officials who presumably still regretted the passing of sail.

Then, in an audacious publicity stunt, Parsons's *Turbinia* turned up unannounced at the Navy Review for the Diamond Jubilee of Queen Victoria at Spithead, on 26 June 1897, in front of the Prince of Wales, foreign dignitaries and Lords of the Admiralty. *Turbinia* raced between the two lines of battleships and cruisers at 34 knots while easily evading a navy picket boat that tried to pursue it. The Admiralty was outraged – but convinced. Within a couple of years, two Parsons steam turbine destroyers had been built, in 1905 the

Admiralty confirmed that all future Royal Navy vessels were to be steam-turbine-powered, and in 1906 the first turbine-powered battleship, HMS *Dreadnought*, was launched. Parsons had been triumphantly vindicated. But Whittle's experience of convincing officialdom of the benefits of the turbine was to be rather different.

It turns out that gas turbines were even older than steam turbines. In 1791 a patent was granted to John Barber, an English coal mine manager and inventor, for the first true gas turbine. His invention had the same elements present in modern-day gas turbines: gas (derived from coal) was burned in a combustion chamber and the hot gases spun a turbine. His engine was designed to power a horseless carriage, and so it could be said that Barber invented a gas-turbine-powered car nearly 200 years before Rover's ran at Le Mans. Contemporary metallurgy just wasn't up to the high temperatures involved in Barber's contraption, and so his idea had to bide its time.

Had anyone else come up with Whittle's idea? The Italians had a configuration called a ducted fan, which involved a piston engine that drove a series of fans within a hollow fuselage, providing thrust. Fuel could be sprayed during the exhaust phase as a form of afterburner, although this didn't appear to add much speed. The Italians would continue work on this idea and flew a ducted-fan aircraft – the Caproni Campini CC.2 – in August 1940. However, the plane was underpowered and unusable. It has been described as a jet, but it wasn't: it was a piston-powered ducted fan. The novel configuration offered no advantage over conventional designs, and development was stopped.

Three years before Whittle's eureka moment, an English engineer named Alan Griffith had published an important paper, *An Aerodynamic Theory of Turbine Design*. He had correctly demonstrated that the poor performance of existing gas turbines was due to a flaw in their design: the use of flat blades meant they were 'flying stalled'. He proposed an aerofoil shape for the blades that would dramatically improve their performance. His paper went on to describe an engine using an axial compressor like a many-bladed fan, and a two-stage turbine, the first stage driving the compressor, the second a power-take-off shaft that would be used to power a propeller. He had already built a working test model by 1928. This early design was a forerunner of the turboprop engine and developed a strong theoretical basis for the jet engine. It was close to Whittle's idea, but it lacked his vital next step of using the excess leftover flow of gases to provide thrust.

Whittle realised that turbojets – or jets, as they are called today – could offer huge advantages. The beauty of a gas turbine compared with a piston engine is that it deals with the air–fuel mixture continuously, instead of in individual batches. The combustion takes place in combustion cans where the pressure is constantly high, not alternating between the 'suck, squeeze, bang, blow' stages in the combustion chambers of a four-stroke piston engine. Turbojets could have a huge compressor, which could process much more air; as we know, a modern aircraft jet engine can suck in two tons of air a second and develop the equivalent of 50,000hp. The largest aircraft piston engines were always restricted in the amount of air they could consume and struggled to make much more than 3,500hp. Due to this enormous power, a jet aircraft can fly at high speeds with low vibration (due to there being no reciprocating pistons), at high altitude, with greater reliability (due to a simpler construction) and with more safety (because jet engines use lower-flammability kerosene fuels). Jets are simpler than pistons engine too – think vacuum cleaner crossed with a paraffin blowlamp.

It has to be pointed out that the idea for gas turbines in aircraft was 'out there', and at least five teams were working on gas turbine engines by 1937. But Whittle seems to have been the first to understand that gas turbine thrust alone would be sufficient to drive an aircraft without the need for a propeller. Griffith had not grasped this and insisted on the use of a propellor.

Whittle discussed the idea with one of his instructors at Wittering, a man who would become intimately involved in the development of the turbojet engine, Flying Officer W. E. P. 'Johnny' Johnson. He had been a patent agent in civilian life and was, at that time, the only officer in the RAF qualified in instrument flying. He took the idea to their commandant, who was impressed and brought the idea to the attention of the Air Ministry. A few days later, Whittle was instructed to report to London.

Chapter 2

An Unappreciated Genius

The times were not propitious for Whittle. On Tuesday 29 October 1929, in the same week as he was trying to interest his superiors in his new idea, Wall Street crashed. This was felt all around the world. It plunged Britain into a depression, signalling the end of the Roaring Twenties. The RAF was now fighting for its life. After the horrific First World War, and under pressure from those advocating peace, Britain was disarming. The country's aircraft industry was fragmented and in vicious competition.

Whittle was advised to meet Alan Griffith of the Turbine Design paper at the Air Ministry's laboratory in South Kensington, where he explained his sketches and calculations.

Griffith would seem to have been an ideal choice to comment on Whittle's ideas. He was eminent in his particular field of metal fatigue, and one of his techniques was still being used until recently. It was the clever use of soap films to study stress problems. He would stretch a soap bubble between strings pegged out in the shape of the metal object under study, and the colouration of the soap film would show the patterns of stress buried in the metal.

Alan Griffith's father was the Victorian socialist science-fiction author George Griffith, who envisaged air warfare between fleets of heavier-than-air flying machines, propelled by invisible means. Writing in 1893, he placed the first flight of such a machine in 1903: the same year as the Wright brothers' first flight. Griffith Senior predicted humans would one day avert a cometary collision with the

Earth, and he also originated the idea of the ten-second countdown for space launches.

Griffith is now much admired by steampunk devotees and proves that if you really want to look into the future, you might do worse than read Victorian science fiction. But could he have ever guessed that his own son might help to invent jet propulsion? Because Alan Griffith had his own ideas on the use of gas turbines in aircraft.

The design Whittle showed him had a large centrifugal compressor at the front and a single-stage turbine at the rear of the engine, the leftover power in the exhaust being used to push the aircraft along. As we have seen, this compressor took the form of a large disc with vanes on it. There were good reasons for the use of a centrifugal compressor. It would be easier to make than a many-bladed axial compressor like a Parsons steam turbine, and it was similar to a piston-engine supercharger, so there was readily available technology to draw upon.

However, Griffith was not encouraging. He pointed out an error in Whittle's calculations and said that the large frontal size of the compressor would make it useless for aircraft. He also said the exhaust itself would provide little thrust.

On the latter two points, he was wrong, and Whittle proved him wrong, but only after years of struggle. What is questionable about Griffith's conclusion is that he already had his own gas-turbine design up his sleeve, which he was to produce later.

One is reminded of Darwin's horror when he saw Wallace's theories on evolution, pre-empting ideas he had been mulling over for decades. Was Griffith trying to squash a possible competitor? At the very least, he should have declared an interest.

Popular history books have not been kind to Alan Griffith, with a common theme of the poor young Frank Whittle being obstructed by a jealous older man, more eminent in his field. Whittle's own autobiography and Golley's *Genesis of the Jet* do everything to support this narrative.

However, in *The Jet Engine Story*,[1] Bill Howard puts an entirely new construction on this simplistic tale. Howard points out that many of the popular accounts of early jet development were written by journalists looking for a good story with a human angle, rather than by professional engineers. Hence the repeated stories of Whittle portrayed as a victim. Howard redresses the balance by pointing out that Griffith's ideas at the Royal Aircraft Establishment (RAE) in 1926 eventually resulted in RAE-derived engines with axial-flow compressors, as used by just about all jet engines today. But it took Whittle's determination to prove his ideas about jet thrust that started the two parallel streams of research and development. Eventually these led to the first-generation British jet engines derived from Whittle's centrifugal compressors, soon followed by RAE/Griffith-derived engines with axial-flow compressors. If only Alan Griffith had a more generous mind.

Crestfallen, Whittle had gone back to Wittering, where he revised his calculations and discovered another mistake that counteracted the original one. His jet engine *would* work! A few days later, he received a letter from the Air Ministry, which said, as the engine was a form of gas turbine, development was impractical because materials did not exist capable of withstanding the combination of high stresses and temperatures expected. As such, they were correct. As we have seen, the piston engine experiences intermittent high heat only during one of the four strokes, whereas the turbine engine must cope with constant high temperature in the combustion stage.

Anyone who has worked with gas-welding torches and pieces of metal knows that the way to melt metal is to hold the flame on the

---

1 Howard, Bill, *The Jet Engine Story,* Farnborough Air Sciences Trust, Farnborough (2019)

workpiece continually, not play the flame on and off the metal. It will glow red, then white, then collapse in a molten puddle. This is what gas turbines and turbojets do to their exhaust turbine blades. The high temperature and pressure requirements of the combustion stage of the turbojet engine would prove to be problematic during its evolution. However, huge improvements in materials were expected soon. Whittle remarked in his autobiography, 'I agree that at the end of 1929 it was before its time, but only by a very few years.'

It was the piston aero engine that drove the improvement in materials. Rolls-Royce was having trouble with red-hot exhaust valves burning out and was experimenting with austenitic nickel-chrome steels – what we would call stainless steel. Just around the corner was a family of nickel-based high-temperature superalloys. These Nimonic alloys would eventually be used for Whittle's turbine blades.

Despite the official rejection, Johnny Johnson urged Whittle to patent his invention and offered help in drawing it up:

> In accordance with regulations, the Air Ministry was formally notified of my action, but I was informed that the Ministry had no official interest in the patent. There was no suggestion that it should be put on the 'secret list'. This meant that after the complete specification was filed and accepted, a little over eighteen months later, the invention was published throughout the world.[2]

Was this a good idea? Whittle's options were either to patent his idea or to keep it as a trade secret. To be a patent, it had to meet the legal requirements of involving an inventive step (or be 'non-obvious' in US law), it had to be novel, and it had to be susceptible of industrial application (or be 'useful' in US law). So, in all respects, the idea was patentable, like Alexander Graham Bell's 1876 patent for the

---

2 Whittle, Frank, *Jet*, Muller, London (1953)

telephone, United States Patent No. 174,465, which became the most valuable patent in history.

Or he could press on with development, trying to keep it a secret and not disclosing his idea to the world: a trade secret. In 1886, ten years after Bell's telephone, Dr John Pemberton created the world's most famous trade secret: the Coca-Cola formula. It contained a small amount of cocaine from coca leaves, which enhanced the tonic qualities. The formula is known as Merchandise 7X, and the company still keeps the secret recipe in a vault in Atlanta. The syrup (which also contains phosphoric acid) is combined with carbonated water by bottlers, and no single supplier of ingredients has the full recipe. Pemberton could have patented the formula for the syrup, but that would have given the company only 20 years of exclusivity rights.

Pregnant mares' urine was the secret ingredient of Premarin, a hormone replacement therapy drug used for the symptoms of menopause. The manufacturer, Wyeth, was the sole supplier of Premarin (and you would think the clue was in the name). Patents were issued on the drug in the 1940s but, long after they had expired, there were still no generic competitors on the market. No rival had succeeded at copying the extraction process, which Wyeth had not patented but kept as a trade secret. All that was known is that there was a good chance of being kicked.

There is a danger in trying to keep a trade secret. It's perfectly possible to reverse engineer, copy and then patent someone else's trade secret. A patent lasts only 20 years, but during that period, the protection is far stronger: independent invention is no defence in a patent suit. However, in wartime (and in contemporary China), protection around intellectual property tends to be ignored.

Filed on 16 January 1930, British Patent No. 347,206 was the first in the world to describe a practical turbojet.

Johnny Johnson is something of an unsung hero in Whittle's story. Not only did he support and advise him as a knowledgeable patent agent, I suspect he also helped to add a feature to the patent drawing, which scuppered rival plans. At the front of the engine, inside the annular inlet, there was drawn a two-stage axial compressor which never appeared on the first Whittle unit to be built. In his patent, Whittle therefore cleverly hedged his bets and describes an engine with two axial-compressor stages and one centrifugal, thus anticipating both routes forward. This feature apparently frustrated Hans von Ohain when he saw it, and this forced him to change his own design.

After the two-stage axial compressor, the incoming air was led to a single-stage centrifugal compressor rather like the supercharger discs we have seen, then the compressed air was let into a series of combustion chambers into which fuel was injected. The burning gases were then guided to an axial-flow exhaust turbine (something like Parsons') with two stages or rows of blades mounted on a rotating disc. There was a ring of stationary stator blades between the rotor stages. This disc was mounted on the shaft, the other, forward, end of which drove the compressor. Finally, the hot gas expanded and accelerated through a divergent nozzle ring. The gases rushing out pushed the engine along: thrust.

Whittle might have now had a patent, but he had no money. He had married his fiancée, a middle-class young woman named Dorothy Mary Lee, and they now had a child. He was a serving RAF officer, and he desperately tried to interest commercial firms in his idea. Johnny Johnson arranged a meeting with the steam turbine company British Thomson-Houston (BTH), whose chief turbine engineer agreed with Whittle's theories. However, BTH did not want to spend the £60,000 (£4.1m today) it thought it would cost to develop. The jet became a joke amongst Whittle's fellow RAF officers; it was dubbed 'Whittle's Flaming Touch-hole' and he was often greeted with, 'Well, how's the old flaming touch-hole getting on?'

He realised that he would have to convince sceptics of the value of his compressor ideas before he could hope to get them to accept the complete engine:

> The turbo-jet engine proposal required a compressor having a pressure ratio of the order of 4:1 and an efficiency of at least 75% while the best aero-engine supercharger of that time (that of the Rolls-Royce 'R' engine which powered the Supermarine S.6 which won the Schneider Trophy) had a pressure ratio of just under 2:1 and an efficiency of 62%. Unless a very great improvement on these figures could be obtained there was little hope for the success of the turbo-jet engine.[3]

The Air Ministry dismissed Whittle's jet engine and told him that development difficulties were too great. In 1934, the Secretary of State for Air wrote:

> Scientific investigation into the possibilities of jet propulsion has given no indication that this method can be a serious competitor to the airscrew engine combination... We do not consider that we should be justified in spending any time or money on it.[4]

These were perhaps the most ill-advised words ever written in aviation history.

The man who wrote them, The Most Honourable Charles Vane-Tempest-Stewart 7th Marquess of Londonderry KG MVO PC, was a leading member of the Anglo-German Fellowship. He was also an anti-Semite and made visits to Hitler, Hess, Göring, Himmler, von Papen and other senior Nazis. This attracted, for him,

---

3 Whittle, Frank, *Jet*, Muller, London (1953)

4 Taylor, Douglas R., *Boxkite to Jet: The Remarkable Career of Frank B. Halford*, Rolls-Royce Heritage Trust (1999)

the popular nickname of 'The Londonderry Herr' (a play on the title of a popular song). When closeted with Hitler in October 1936, the Führer indiscreetly revealed to Londonderry his plans to invade Czechoslovakia and Poland. To his credit, Londonderry passed this privileged information on to Lord Halifax on 24 December 1936, but it had little effect. Londonderry was the second cousin of Winston Churchill, who thought him devoid of talent and removed him from office.[5]

I do not for a moment suggest that Londonderry sought to weaken the RAF; in fact, he presided over the planning for fighters such as the Hurricane and Spitfire, and he supported radar research. As an enthusiast for Nazi Germany though, it is possible that he underestimated the German threat. Now, with perfect hindsight, the Air Ministry might have thought to place the jet engine on the secret list and then persuade a company like Rolls-Royce to work with Whittle and Griffith to build a flying protype. And perfect hindsight is just one reason why aviation historians like me are so annoying.

If the Air Ministry had been remiss in taking Whittle's jet engine seriously, at least the RAF recognised his talent. In an unusual move, it sent him to Cambridge University to take the Mechanical Science Tripos, where his studies confirmed his belief that his engine would work. But then:

> In January 1935, my 1930 jet engine patent became due for renewal. For this a fee of £5 was required. Shortly before the renewal date

5 Penguin published Londonderry's book, *Ourselves and Germany*, in 1938. Unusually, the subtitle on the cover contained the entire burden of the book: *Ourselves and Germany: Should Britain regard Germany as a potential enemy or seek her friendship? Lord Londonderry thinks we should adopt a policy of friendship with Hitler and a better understanding of Germany's aims.*

I received an official letter from a branch of the Contracts Directorate of the Air Ministry reminding me that the renewal date was near and adding that there was no intention of paying the renewal fee out of official funds.

Whittle could ill afford the money (some £360 today); he had medical expenses arising out of the illness of his elder son (there was no National Health Service), and the birth of a second child was expected soon. Unfortunately, he let the patent lapse.

Meanwhile, others were thinking about gas turbines for aircraft. In 1921, a Frenchman, Maxime Guillaume, patented[6] an axial-flow turbine engine. It used multiple stages in both the compressor and turbine, combined with a single large combustion chamber, but it was never built, as it would have required considerable advances over the state of the art in compressors.

There were others working on the idea.

One of the tragedies of the Cold War is that Eastern European pioneers did not get the recognition they deserve. Who knows today about the brilliant Tatra cars from Czechoslovakia that built the fortunes of Volkswagen? And who has heard of the Hungarian physicist and mechanical engineer György Jendrassik? Because in 1937, Jendrassik started to build the CS-1, a gas turbine turboprop engine, which he produced and tested at the Ganz works in Budapest. His forward-looking design was an axial-flow design with a 15-stage compressor and a seven-stage turbine that would then drive a propeller. However, it did not use jet thrust. It was more sophisticated than Griffith's turboprop idea, and it foreshadowed many modern designs. With a predicted power output of 1,000bhp

---

6 French patent no. 534,801, filed 3 May 1921, issued 13 January 1922.

at 13,500rpm, the Cs-1 aroused interest in the Hungarian aircraft industry. Construction began on a twin-engined fighter-bomber, the Varga RMI-1 X/H. Jendrassik's engine's first bench run took place in 1940, making it the world's first turboprop engine. However, combustion problems were experienced, which limited the output to around 400bhp. This was not going to be an easy technology to master.

Now we come to the young man who first got the jet engine flying. A 19-year-old German student, Hans Joachim Pabst von Ohain, was studying physics and aerodynamics at the university of Göttingen. In 1933, while still a student, he came up with what he called 'an engine that did not require a propeller'. This was three years after Whittle's patent was published in specialist magazines in Germany.

The 1919 Treaty of Versailles had forbidden Germany from building an air force. Disarmament was spreading amongst other countries in the peace-loving 1920s, with nations disbanding their air forces and cancelling contracts for military aircraft. Research and development in Britain was left to private contractors such as Rolls-Royce and de Havilland, which had to privately fund their Merlin engines and Mosquito fighter-bombers.

As a result of the Versailles treaty, Germany decided to conduct aeronautical research in secret. As for the nation's aircraft production, the aircraft manufacturer Fokker had based itself in the Netherlands and was secretly selling its products to Germany, and Heinkel sold aircraft through Sweden back to Germany. Military pilots were covertly trained in Russia and Italy.

When the Nazis took power in 1933, they started overt rearmament. Air power was crucial to their ambitions, and on 1 May of that year, Hitler's deputy, Hermann Göring, was appointed Reichskommisar of a new air ministry: the Reichsluftfahrtministerium (RLM), which started to pour funds into aeronautical research. The flood of money principally benefitted the German aircraft manufacturing firms Heinkel, Messerschmitt, Junkers and Dornier, and the engine companies BMW, Daimler-Benz, and the engine division of Junkers.

The RLM trained a cohort of aeronautical engineers between the wars. Its machinists were trained in Machinenbauschulen, while its engineers attended the Technische Hochschulen where they studied physics, chemistry and materials engineering. And so, Germany's technical superiority by the beginning of the Second World War was no accident: the country had been secretly preparing for another war for 20 years. By 1938, their military aircraft were the best in the world.

And this time they were determined to win.

Ironically, it was the Versailles Treaty's ban on conventional aero engine development that stimulated German interest in unconventional gas turbines for aircraft. Furthermore, it could be argued that the Germans actually benefitted from Versailles and the limitations it imposed on aviation research. The Luftwaffe had to start from a blank sheet of paper, so there was less conventional thinking, whereas in Britain there was great resistance to the move away from biplanes because older officers remembered their glory days of dogfights over the trenches.

One of the top German schools was the Aerodynamics Research Institute at the university of Göttingen. Ludwig Prandtl, a theoretical physicist, lectured there to the young doctoral student Hans Joachim Pabst von Ohain, who said he had an idea for an engine.

Ohain's design had a two-stage compressor, with one axial-compressor stage followed by a single-sided centrifugal-compressor stage, an annular combustion chamber and a radial turbine. The axial compressor and turbine were placed close together, back-to-back, with the flame cans wrapped around the outside of the assembly. It looked rather like Whittle's design.

He now needed to test his idea. While studying at Göttingen, Ohain used to take his sports car to be serviced at a local garage, Bartles and Becker. There, he met a mechanic, Max Hahn, who built him a working model of his design for 1,000 Reichsmarks (around £5,000 today).

This first engine was put together out of sheet steel and nuts and bolts, as Hahn wasn't too confident about welding. The finished model was even larger in diameter than Whittle's fully working engine of 1937, but it was shorter along its thrust axis. Ohain took the model to the university for testing but, like Whittle, ran into problems with combustion stability. At first it wouldn't run. He could not persuade the flames to stay inside the flame cans, and burning fuel would blow through the turbine, sending flames shooting out into the exhaust.

Eventually, he persuaded the flames to stay in the combustion chambers, and the engine speeded up. It wasn't going to be powerful enough to fly, but it did run, and it was self-sustaining: it kept running when the starter motor spinning the compressor was disconnected.

Although the project was a failure as a turbojet power plant, Ohain's theoretical analysis and his rudimentary prototype proved the validity of turbojet technology. On 10 November 1935, he filed patent 317/38 for his version of a turbojet engine, 'Process and Apparatus for Producing Airstreams for Propelling Airplanes'.

Had Ohain copied Whittle? The question still causes furious debate amongst aero historians, with shrill accusations of nationalism from both sides. It has been claimed on one side that Ohain was unaware of Whittle's work because, initially, Ohain said that was the case, and when they met after the war, Whittle, trusting as always, believed him.

However, the Whittle patent was in German libraries and magazines, and Whittle's son had suspicions that Ohain had seen it.

Years later, Ohain admitted in his biography that this was indeed the case. He said that his patent attorney had found the Whittle patent while the Ohain patents were being formulated and had drawn it to his attention.

Ohain admitted that he had studied Whittle's patent for a gas turbine engine, as would only be expected of an academic working in the field. Then he filed his own patent, which was registered in 1935, some five years after Whittle's. In his biography, Ohain criticised Whittles design:

> When I saw Whittle's patent, I was almost convinced that it had something to do with boundary layer suction combinations. It had a two-flow, dual entrance flow radial flow compressor that looked monstrous from an engine point of view. Its flow reversal looked to us to be an undesirable thing, but it turned out that it wasn't so bad after all although it gave some minor instability problems… Our patent claims had to be narrowed in comparison to Whittle's because Whittle showed certain things.

Here his 'certain things' possibly refers to Whittle's broad claims of both axial and centrifugal turbines. He then justified his knowledge of Whittle's work by writing: 'We felt that it looked like a patent of an idea… We thought that it was not seriously being worked on.'[7]

When he initially said he was unaware of Whittle's work, perhaps Ohain meant he was unaware of Whittle's current successful tests at Lutterworth. Whatever the truth of it, the two jet pioneers became friends after the war and Whittle always accepted Ohain's version that he was unaware of his predecessor's work.

It would be most surprising if Ohain had been oblivious to Whittle's ideas. The idea of gas turbines for aircraft was clearly

---

7 Quoted in Conner, Margaret, *Hans von Ohain: Elegance in Flight*, American Institute for Aeronautics and Astronautics, Inc., University of Michigan (2001)

floating around in the international engineering community, just as the idea of piston engines was in the air 35 years previously. Foreign patents were carefully studied. Whittle had added the inventive step of using excess thrust to propel the aircraft. Someone had to be the first to get the jet engine flying. But pioneers rarely get any credit for their work, because no one knows what they're doing.

As Frank Whittle had discovered, the development of this exotic new technology was going to be enormously expensive, too expensive for an individual. However, the young Ohain had far more fertile ground to sow his idea in: Germany was more inclined to spend money on new technology than Britain. And it is hard to believe that Hans von Ohain's different treatment didn't have something to do with his different circumstances. Unlike Whittle, he came from an aristocratic family; his father was a German Army officer, and he had the money to drive a sports car as a student. He was able to spend £5,000 for the project. But oddly enough, it wasn't the German RLM that picked up the turbojet idea, it was Dr Ernst Heinkel.

Ohain's adviser, Professor Pohl, was so impressed with young Ohain's theoretical analysis and running model that he said, 'This is sound… Tell me what company you would like to work with on your engine and I'll write a letter of recommendation for you to its director.'

Ohain chose the Heinkel Company because, he replied, 'Heinkel has the reputation of being a little bit crazy. I am considered crazy too so maybe we'll make a good team.'

Ohain had taken a flight in a Junkers Ju 52 and was deafened by the three BMW radial engines. He remarked, like Whittle, that piston engines were far too rough, noisy and dirty: 'I thought the reciprocating engine was a horrible monster and that the propeller lacked elegance.

I thought that it should be possible to drive an airplane fast and smoothly through the air like a very fantastically fast glider.'[8]

Pohl wrote to Heinkel, who was always open to new ideas, and he was sufficiently interested to write back, inviting Ohain to his home. This meeting was on 18 March 1936.

Unlike their British cousins, German industrialists knew a good thing when they saw it, and Heinkel immediately saw the potential of the idea. In April 1936, Ohain and Hahn began working for the company at the Marienehe airfield outside Rostock, in Warnemünde. Ohain was given the encouragement and the RLM's funds to develop his engine. The German government didn't even know about the Heinkel project because his rearmament contracts enabled money to be funnelled into his company's research and development. He used the money well, gathering some of the finest young aeronautical engineers and designers together to work at his aircraft manufacturing firm and hiring the brilliant Günter twins.

The twin brothers, Siegfried and Walter, were radical designers who were going to design some of the best aircraft for the company, introducing Prandtl's elliptical wing (used on the Spitfire) and the retracting undercarriage. Their He 70 airliner was purchased by Lufthansa, which named it the Heinkel-Blitz ('lightning') and instituted 'blitz' air routes between Berlin, Cologne, Hamburg and Frankfurt. The twins had both fought in the First World War and had both been taken prisoner by the British. The Günter twins would develop the world's first turbojet airframe with Ohain's engine, and the world's first rocket-powered airframe.

Dr Heinkel was a notably foresighted designer, always fascinated by high speed. Even after 1942, when he was forced to sell his aircraft company to Hermann Göring, he continued to

---

8 From 'Interviews with Hans von Ohain' in *Interviews with German Contributors to Aviation History*, Joseph Ermenc (ed.), Kappus, Westport, CT (1990)

design, and after the war, when even denied permission to build aircraft, he still came up with the memorable Heinkel Kabine bubble-car, a 174cc, 9.2hp scooter-engined three-wheeled car with a top speed of 54mph (87km/h).[9]

Meanwhile, in 1935, Whittle had been approached by RAF colleagues who had managed to raise money for his ideas:

> I gave up hope of ever getting the idea to the practical stage, but continued to do paperwork at intervals, until, in May 1935, when I was at Cambridge as an engineer officer taking the Tripos course, I was approached by two ex-RAF officers (Mr R. D. Williams and Mr J. C. B. Tinling), who suggested that they should try to get something started. Although I had allowed the original patent to lapse through failure to pay the renewal fee, and although I regarded them as extremely optimistic, I agreed to co-operate. I thought that there was just a bare chance that something might come of it.[10]

With Williams and Tinling on board, Whittle was able to come to an arrangement with investment bankers O. T. Falk and Partners, leading to the formation of Power Jets in 1936. Whittle now pressed on with developments, but once again he came across his nemesis, Alan Griffith. The Aeronautical Research Committee (ARC) had asked for an evaluation of Whittle's design. The report was passed on to Griffith for comment, but it was not received back until March 1937, by which point Whittle's design had progressed. Griffith had already started construction of his own turbine engine

---

9 This Heinkel featured a reverse gear, unlike some bubble cars with a single front door. If you parked too close to a wall in one of those cars, you were unable to open the door and unable to reverse, remaining trapped inside until help arrived.

10 Whittle, Frank, *Jet*, Muller, London (1953)

design and, perhaps to avoid tainting his own efforts, he returned a grudgingly positive review. However, he remained sceptical of some features, notably the use of jet thrust: the whole point of Whittle's inventive step.

His conclusion was:

> [...] in its present form the proposed jet propulsion system cannot compete with the conventional power plant in any case where economical flight is demanded (e.g., the transport of the maximum percentage of useful load over given distance). It is of value only for special purposes such as the attainment of high speed or high altitude for a short time in cases where take-off requirements are not stringent.

The Engine Sub-Committee of ARC studied Griffith's report and decided to fund his effort instead.

With grim determination, and now some money behind him, Whittle pressed on. His planned engine was to be a simple jet propulsion gas turbine having a single-stage double-sided centrifugal compressor with bilateral intakes, driven by a directly coupled single-stage turbine. Combustion was to take place in a single chamber through which the gases passed from the compressor to the turbine. Then there was a single propelling nozzle. It had a pressure ratio of 4.4:1, a turbine entry temperature of 780°C and a design thrust of 1,389lb.[11]

Whittle wrote:

> We were going beyond all previous engineering experience in each of the major organs. We were aiming at a pressure ratio of about 4/1 in a single-stage centrifugal blower when at the time, so far as we knew, a ratio of 2/1 had not been exceeded. We were aiming at a breathing

---

11 Ruffles, Philip, *The History of the Rolls-Royce RB211 Turbofan Engine*, Rolls-Royce Heritage Trust, Derby (2014)

capacity in proportion to size substantially greater than had previously been attempted. The combustion intensity we aimed to achieve was far beyond anything previously attempted. Finally, we had to get over 3,000 SHP (shaft horsepower) out of a single-stage turbine wheel of about 16 in outside diameter, and to do it with high efficiency.

When reading Whittle's autobiography, *Jet*, you are struck by the endless troubles he faced from the companies he worked with. British Thomson-Houston, which was a heavy steam-turbine manufacturer, was contracted to build his engine but seemed sceptical of his design. It offered little help, and appeared to resent it when Whittle discovered that, all along, it had been making its own steam turbine blades in the wrong shape! In his typically analytic way, he had deduced that there was a pressure difference between the inside and the outside of the annular ring of gases leaving the nozzle ring. Steam or turbojet gases would behave in much the same way. Whittle realised that the change of angle or 'twist' from root to tip of the turbine blades ought to be twice that provided in the BTH turbine blade design.

Instead of being grateful that the young inventor had pointed out a way of improving its main product, resentment grew, especially when Power Jets filed a patent for the 'twist'. At one point, Whittle even suspected sabotage from persons unknown when loose nuts were found deposited inside his turbojet engine, damaging it on start-up. In the end, BTH decided it was not really interested in making jet engines.

One wonders if there was something about Whittle's personality that irritated the people he came across. He had been told that he was a prima donna by no less than Lord Tedder. And a friend of mine was Whittle's doctor when he was admitted to a RAF hospital. He described Whittle as 'a difficult person', and several of his medical staff complained of his rudeness and eventually refused to enter his room. Whittle's tragedy was that he was a brilliant and a driven individual and this made him impatient with mediocrity. He naïvely underestimated the duplicity of his competitors, and he failed to understand that the British government was never going to fund an entirely new form of propulsion. He was, therefore, doomed to disappointment and disillusionment.

Whittle's jet engine ran up for the first time on 12 April 1937, five months before his unknown rival Ohain's device. He was apprehensive, excited even, but it turned out to be rather more exciting than he expected.

> For a second or two the speed of the engine increased slowly and then, with a rising shriek like an air-raid siren, the speed began to rise rapidly, and large patches of red heat became visible on the combustion chamber casing. The engine was obviously out of control… I screwed down the control valve immediately, but this had no effect and the speed continued to rise…[12]

Despite Whittle's desperate attempts to slow his engine, it started to overspeed. If a turbine 'runs away' like this, it means that it uncontrollably spins up to dangerous revolutions. The poor overheated turbine blades, clinging on by their roots, become heavier and heavier as they whirl around. The poet W. B. Yeats might almost have been writing about jet engines:

Turning and turning in the widening gyre
The falcon cannot hear the falconer;
Things fall apart; the centre cannot hold;
Mere anarchy is loosed upon the world…[13]

---

12 Whittle, Frank, *Jet*, Muller, London (1953)

13 Yeats, W. B., 'The Second Coming' (1919)

Anarchy would indeed be loosed upon the world as an overspeeding turbine could suddenly explode, flinging dozens of knife-like turbine blades at supersonic speed through casings, walls and people. Steam turbines had burst in the same factory with fatal consequences. The BTH personnel standing around Whittle's engine, knowing what could happen, ran for safety, some diving into nearby steam turbine casings. Whittle stuck to his controls: '… fortunately, the acceleration ceased at about 8,000 rpm and slowly the revs dropped again. Needless to say, this incident did not do my nervous system any good at all. I have rarely been so frightened!'

The team tried to start the jet engine again the next day, but this time, there were sheets of flame 'leaping and dancing above in mid-air above the engine' and 'sheets of flame belched from the jet pipe' as the engine ran out of control again.

All this is reminiscent of starting an old-fashioned Primus camping cooker. If the jet of pressurised fuel fails to vaporise and squirts out as neat liquid, the result is a huge sheet of flame, which usually ignites you, your clothing and the tent you're sitting in.

The cause of the Whittle jet-engine fires turned out to be a faulty fuel system, which had allowed the fuel to puddle in the bottom of the combustion chamber, driving engine to high revs even when the supply was turned off. On the fourth run, this 'rough beast' of an engine behaved better, consoling itself with merely melting its ignition cables.

At last, Whittle had a running, powerful jet engine. Further testing and development meant that he could measure the power being developed. The engine was run up to 13,000rpm and produced 480lb thrust. Because of the ramshackle shed he was working in, it was dangerous to take any readings at all: 'It was rather a hazardous business getting a thrust reading, because it was necessary to go past the engine and note the reading of the spring balance which linked the test truck to a post embedded on the test room floor.'[14]

---

14 Whittle, Frank, *Jet*, Muller, London (1953)

It is worth pointing out at this point that measurements of power are different between piston engines and jet engines. Engines such as the Supermarine Spitfire's Rolls-Royce Merlin have their horsepower measured at the shaft that drives the propeller. It is a calculation of twisting power (i.e., torque) multiplied by time: revolutions per minute. Whittle's jet engine had its power measured by the direct push it would give to the aircraft: the thrust. Thrust is force, whereas horsepower is a measure of work done, so the two measures are hard to compare. But for a small fighter plane flying at 350mph (560km/h), 1,000lb thrust is roughly equivalent to 1,500hp.

Whittle had started with a simple centrifugal compressor, which, although having a larger frontal area and lower efficiency than the multistage axial compressor, was already well-proven in piston engine superchargers. His biggest problem was with the combustion chamber, and in the WU3, his third design, he substituted the single combustion chamber with a new design:

I abandoned the idea of trying to make a satisfactory arrangement with a single combustion chamber and produced a layout with ten small combustion chambers (to conform with the then existing ten discharge ducts from the compressor). This made a much more compact and lighter engine... I hit upon the idea of 'inter-connecting tubes' by which means each combustion chamber would light up from its neighbour... We did not realise it at the time but the apparatus we used was very similar to the successful combustion system which was developed much later after many heart-breaking disappointments... some slight adjustment might have made all the difference and might have saved three years.[15]

---

15 Whittle, *Jet*, p. 79

Fighting deteriorating health, passive resistance, resentment and outright intellectual theft, Frank Whittle persevered until his engine was ready for an airframe that would test his idea in the air.

Back in Germany, by the mid-1930s, ferocious competition had arisen between the aircraft companies Heinkel and Messerschmitt. They were both building aircraft that contained the latest technology, and each took turns in being the fastest in the world. There would be intense rivalry between the two companies over contracts over the course of the war, and both would eventually develop turbojet aircraft for the Luftwaffe.

Dr Ernst Heinkel was, therefore, motivated to develop a new engine that could beat Messerschmitt hands down. He was also testing a rocket-propelled aircraft, his He 176. By the time Whittle had run his jet engine for the first time, Heinkel's team had started to develop the Heinkel HeS 1 turbojet engine. This engine burned gaseous hydrogen for fuel and would not have worked otherwise, but it met the theoretical expectations of performance and convinced Heinkel that turbojet technology would work. The next goal was to get an engine running on liquid fuel. As Whittle was finding, the difficulty was the combustion stage: it would involve enormous pressures and sustained high temperatures. Finding materials that could withstand these conditions would continue to be a problem. The Heinkel team persevered, and a flight-ready engine was built, the HeS 3b, by early spring 1939.

Meanwhile, the Günter twins had designed a tiny shoulder-winged aeroplane to test the engine, the He 178. It was built around the physical measurements of the chief Heinkel test pilot, Erich Warsitz. At this point, Heinkel showed Hitler his rocket plane and the turbojet plane. The hydrogen peroxide monopropellant rocket flew successfully, and Ohain explained the new jet technology to the Führer. It was not until now that the RLM was informed of Heinkel's secret jet technology. He promised them a flight by the end of July.

It turned out that other German turbojet projects were under development, but Heinkel's led the field. Ohain's story had spread through the industry. There were further delays, and so it was decided that the first flight would be fuelled by compressed hydrogen. This was because of the time pressure imposed by Heinkel's promise, but they knew a liquid-fuelled engine was not far off.

And so it was that on 27 August 1939, Erich Warsitz took off from the airfield at Marienehe, ushering in a new era of jet-powered flight. 'The hideous wail of the engine was music to our ears,' said Heinkel.

Just five days later Germany marched into Poland, marking the start of the Second World War.

Chapter 3

Any Sufficiently Advanced Technology is Indistinguishable from Magic[1]

'I invent nothing,' remarked Henry Royce dryly. 'Inventors go broke.' Although not entirely true – Royce invented a crankshaft vibration damper for his 1903 six-cylindered engine – he was right as regards the struggling Frank Whittle. In 1941, the Rover car company was chosen to replace BTH in making Whittle's jet engines, but he soon discovered the company was copying his designs, removing all references to Power Jets and the origins of the drawings. It was then making changes to his design and building its own versions of the jet engine in secret. He realised the company was trying to 'borrow' his technology for a commercial advantage after the war.[2] His patents were almost worthless as he was still a serving RAF officer, and the Air Ministry would get

---

1 Arthur C. Clarke's Third Law from *Profiles of the Future*, Pan, London (1973)

2 A whole book could be written about Whittle's tragic struggles with businessmen and the Air Ministry, and it has been: Duncan Campbell-Smith's *Jet Man: The Making and Breaking of Frank Whittle, Genius of the Jet Revolution*, Head of Zeus, London (2020). Recommended.

first use of any jet development before commercial firms. Whittle became increasingly frustrated and continually harassed the Rover development team.

However, Whittle's engine was running better and better, and in December 1940 Air Chief Marshall Sir Hugh Dowding came to observe the top-secret new engine. The visit was memorable:

> While the experimental engine was running, I took him round the outside of the test houses. It was impossible to speak because of the noise. As we stood about three yards away from where the jet nozzle protruded through an aperture in the test-house wall, I pointed to the nozzle, meaning to imply 'That's the business end of the engine.' He misunderstood my gesture and, before I could stop him, walked rapidly in the direction indicated. Suddenly a mighty invisible force wrenched open his raincoat and sent him staggering across the concrete – his 'brass hat' rolling away on to the grass.[3]

Although Whittle was appalled, Dowding was clearly impressed by the power of the engine and afterwards managed to joke about the incident. Whittle told him how the cows in the next field used to assemble behind a hedge in the pleasant warmth of the jet.

Whether it was because of this demonstration or not, the ARC now strongly recommended that the Air Ministry should press ahead with gas turbine developments both from Whittle's Power Jets and from the RAE, who were following Griffith's designs. The Ministry therefore commissioned an experimental airframe from the Gloster aircraft company. This was just after the outbreak of war in September 1939.

When Whittle demonstrated that his jet would provide propulsion, Alan Griffith was forced to revise his opinion. He quickly redesigned

---

3 Whittle, Frank, *Jet*, Muller, London (1953)

his axial engine in early 1940 and had it built by the steam turbine makers Metropolitan-Vickers (Metrovick). This resulted in the Metrovick F.2, which ran for the first time later that year.

This new engine was more advanced than Whittle's; it had a nine-stage axial compressor, annular combustor and a two-stage turbine, and it was smaller and more powerful than the Whittle engine. The jet power race had begun.

Eighteen more months of hard work on Whittle's engine and the Gloster airframe followed. Eventually, Whittle's W1 engine was fitted and with the Gloster chief test pilot, Flight Lieutenant Gerry Sayer, at the controls, the Gloster E.28 took off by accident for around 200 yards during taxiing trials from the company's airfield at Brockworth, Gloucestershire.

Like the Wright brother's first hop, aero historians don't seem to count this first flight. The Gloster E.28 transferred to RAF Cranwell, which had a longer runway, and on 15 May 1941, Sayer flew the aircraft under jet power for the first time from Cranwell, near Sleaford in Lincolnshire, in a flight lasting 17 minutes. A maximum speed of 370mph (600km/h). was attained, in level flight at 25,000ft with 17,000 turbine revolutions per minute, compared to the maximum of 3,000rpm achievable by piston aero engines. The jet aircraft was already faster than a Spitfire, and it had a great deal more to give. It was a particularly sweet moment for Whittle, as RAF Cranwell was his alma mater, and, perhaps for that short time, all his struggles might have seemed worth it.

One striking feature of this flight is that the Gloster E.28 was taxied into a hanger and the Power Jet crew left it there without inspection. They knew by the smooth note of the engine that all was well. This dumbfounded the Gloster crew, who were used to piston-engine mechanics tearing off the cowlings after every test flight. This was an early indication that jet engines were going to need far less attention than the old piston engines.

Test pilot Gerry Sayer disappeared during a test flight in a Hawker Typhoon on 21 October 1942, presumed killed in a collision, so his assistant, Michael Daunt, took over testing of the E.28/39. Test pilots have always had a high death rate. The second prototype was destroyed in a high-altitude flight on 30 July 1943 when the wrong type of grease in the controls led to the freezing-up of an aileron. The test pilot suffered frostbite after he baled out. But the success of the E.28/39 had proved the concept of jet-powered flight to the British Air Ministry, and Gloster pressed ahead with designs for a production fighter aircraft. Due to the limited thrust available from early jet engines, it was decided that the next production aircraft would be powered by a pair of turbojet engines. On 7 February 1941, Gloster received an order for 12 prototypes under Air Ministry Specification F9/40. It would become the Gloster Meteor.

Meanwhile, Britain's American allies had got wind of their new power plant for aircraft. General Henry Harley 'Hap' Arnold was the commander of the United States Army Air Force (USAAF), as it was called in 1941. He was a keen pilot with an interest in new technology. When he visited England in April 1941, his nation was officially still neutral, France had been defeated, Stalin and Hitler were still following policies of non-aggression under the terms of the Molotov-Ribbentrop Pact, and Pearl Harbor was seven months away. Britain stood alone against Hitler and the Axis powers.

Arnold was visiting in connection with the Lease-Lend agreement, but what really impressed him was Whittle's revolutionary new jet engine in the Gloster E.28. Arnold immediately saw the value of the new technology and asked the British to share their knowledge and to supply America's first jet engines.

One relevant technology the Americans had mastered was turbosuperchargers for conventional piston aeroengines. We now call these devices turbochargers, and around 50 per cent of modern motorcars have them fitted. Like a directly driven supercharger, they force extra air into the engine. They have a striking similarity to turbojets, having an axial compressor that pressurises the incoming air and an axial exhaust turbine that drives the compressor on the same shaft. This, like the turbo housing itself, has to be made of a metal that can withstand the intense heat. Of course, instead of a combustion chamber like Whittle's new jet, they used a piston engine; the pressurised air was forced into the combustion chambers and the hot exhaust gases supplied the exhaust turbine. General Arnold realised that what Frank Whittle had done was to throw away the piston engine and replace it with a combustion chamber. Huge volumes of fuel and air could therefore be burned without a restrictive, clanking old petrol engine in the way. The result would be power, and lots of it.[4]

The Americans had much of the expertise in metallurgy to make exhaust turbines. What they needed was Frank Whittle's combustion chamber expertise. And so in 1942, he was flown to General Electric's plant in Boston to help it with its Type I Supercharger, as the secret engine was dubbed. Although he suspected that his accommodation was bugged,[5] he was willing to share everything with his country's allies. In fact, all the domestic staff were FBI agents. The Americans were determined to steal their allies' secrets.

---

4 Enterprising mechanics keen to upset their neighbours can, therefore, build a small turbojet out of an ordinary motorcar turbocharger, and this can be done by splicing in a simple combustion chamber between the inlet compressor and the exhaust turbine, as can be seen on YouTube. Noisy but fun.

5 Golley, John, *Genesis of the Jet,* Airlife, Shrewsbury (1996), p. 194

Whittle began to show signs of nervous strain after his years of struggle, and his compatriots at the British Air Commission grew concerned. He clearly needed a rest. They flew him out to California, where he was installed in the Beverley Hills Hotel and taken out to parties, where he met actors, ruined his watch in a 3am swimming pool party and learned to dance the rumba. Frankie really went to Hollywood: 'My diary of the remainder of my stay in California is very sketchy, and my memories of it are somewhat disjointed. I found the treatment much more congenial than the Sun, rest etc., so much so that after ten days I felt the urge to return to work.'

Returning to General Electric, he found it was already planning for a production rate of 1,000 engines a month, very different to the way things were back in Austerity Britain.

The Americans had been handed the new jet engine technology on a plate. And the British bent over backwards to assist their allies: in July 1943, one of only two H-1 jet engines available was sent to the United States, where it was fitted to the prototype P-80 Shooting Star, which first flew in January 1944. The Americans accidentally destroyed this engine during ground testing, so the British sent them their only remaining H-1 from the prototype Vampire.

This generosity was not reciprocated after the war, when the United States failed to assist the British, as it might have done.

A curious reader might wonder why the British Air Ministry seemed so slow to seize on the new jet technology. There was a reason for this.

During the First World War, the Ministry of Munitions and the Air Board were panicking about the slow rate of aero-engine production. They exerted great pressure on Rolls-Royce to allow other companies to make their aero piston engines under licence, but the managing director, Claude Johnson, was adamant that Rolls-Royce quality would not be upheld and said the plan would 'yield nothing but mountains of scrap'.

Rolls-Royce was not a mass-production company at that time, and until 1914 it had not made one aero engine, so perhaps Johnson felt protective of the Rolls-Royce reputation. And yet, in the next war, the Merlin was made in the thousands under licence by Packard and Ford. This stubbornness almost led to Rolls-Royce being nationalised and thus would have been forced to let others make its engines. The justifiable frustration being felt by the Ministry officials led to a disastrous decision: to buy another untried aero piston engine.

'The ABC engine, the Dragonfly, was a piece of rampant opportunism that might have lost us the war had the war lasted much longer; for the government were completely seduced by its design and the salesmanship of its designer Granville Bradshaw,'[6] so wrote L. J. K. Setright, the Old Testament prophet of technical journalists. Granville Bradshaw was a considerably better salesman than designer and he managed in early 1918 to persuade Sir William Weir, the Director of Aeronautical Supplies to 'cancel all the others', that is to say stop production of all other well tried and tested engines apart from the Rolls-Royces and the Siddeley Puma. Weir made the disastrous decision to place large orders for the untried Dragonfly, with 11,500 engines having been contracted from 13 suppliers by June 1918.

The Dragonfly looked good, an air-cooled radial with shiny copper-plated fins for supposedly better cooling. But when it finally ran, it weighed 66lb (30kg) more than promised and developed 45 fewer horsepower. It proved to be one of the worst cooled engines ever built, growing so hot that the cylinder heads glowed red and charred the wooden propeller. Worse was to follow: it turned out that it was designed to run at the torsional resonance frequency of its own crankshaft, causing severe vibrations that caused it to break up

6 Setright, L. J. K., *The Power to Fly*, Allen and Unwin (1971)

after just a few hours. Sir William's unwise decision and Bradshaw's cupidity could have lost the British the war in the air.[7]

The head of engine development at the ministry, Major George Bulman, who was now desperately getting the Merlin into production, had opposed the disastrous Dragonfly back in 1917. Now, in the next war, he didn't need another plausible salesman pushing an untried technology onto the RAF.

Another reason is that it was too late. Britain was fighting for its life during the 1940 Battle of Britain; the Rolls-Royce Merlin had only just become reliable, and it had only just started on the remarkable development curve that saw its horsepower double.[8]

The priority was for fighters that worked dependably and that could be developed to stay just ahead of the German fighters. And after the Battle of Britain was won, attention switched to the production of heavy bombers, which would pummel the German homeland and industry. Again, they needed tried and tested engines with good fuel consumption and a long range, not thirsty and untested jets. As Hans von Ohain had said, if the British had fast jet fighters in 1940, it is unlikely Hitler would have ordered an air attack on their nation. One wonders what kind of accommodation might have been reached between the two powers?

The Gloster Meteor was delayed due to the Rover company, which was struggling to manufacture the W.2 engines on schedule. Senior engineer Stanley Hooker of Rolls-Royce had been to

---

7 BMW engineers were seen sketching Granville Bradshaw's ABC motorcycle at the 1919 Motor Cycle show at Olympia. This featured an air-cooled flat twin engine laid across a duplex cradle frame, with a car-style plate clutch and 4-speed gearbox. The ABC failed to sell, but the BMW R32 that appeared in 1923 had all these features and founded a dynasty of BMW bikes that survives today.

8 See Hoyland, *Merlin*.

see Whittle and his new engine. 'What power does it produce?' Ernest Hives, chairman of Rolls-Royce, asked when Hooker returned to Derby. Hooker replied, 'About a 1,000 pounds of thrust.' He went away and did some slide rule work and returned to Hives, 'Do you know what the thrust of a Merlin in a Spitfire is at 350 mph?' he asked. 'No,' said Hives. 'How much?' '1,000 pounds,' said Hooker. Hives at once arranged for an early visit to Lutterworth![9]

Ernest Hives foresaw that, after the war, there would be a glut of Merlin engines for civil transport and therefore a crash in his company's sales. He realised that the new turbojet would provide Rolls-Royce with a huge new market, and so he made his move. He and Hooker took Spencer Wilks of Rover to dinner at the Swan & Royal Hotel, Clitheroe. Hives asked Wilks, 'Why are you playing around with the jet engine? It's not your line of business, you grub around on the ground, and I hear from Hooker that things are going from bad to worse with Whittle.'

Wilks ignored the jibe and replied, 'We can't get on with the fellow at all, and I would like to be shot of the whole business.'

Hives then said, 'I'll tell you what I'll do. You give us this jet job, and I'll give you our tank factory in Nottingham.'[10]

And so the deal was done. Rover could wash its hands of the jet, and Rolls-Royce got world-leading technology for peanuts.

It was the right decision. By January 1943, the running time of the W.2B jet at Rolls-Royce was already 400 hours, ten times what Rover had achieved in the previous month. The engine would be named the Welland, beginning a convention of naming Rolls-Royce turbojet engines after rivers, symbolising the steady flow of power through the engine.

---

9 Golley, *Genesis of the Jet*, p. 178

10 Ibid. The Rolls factory at Barnoldswick gave the jet engines their 'RB' designation.

The Welland is a particularly beautiful river, meandering through eastern England and forming the county border between Rutland and Northamptonshire. Henry Royce's family had owned a mill on the river, at Seaton. Hence the connection.

One of the W.2B engines was fitted into the tail of a testbed Wellington bomber, which retained its usual two piston engines. One day, pulling alongside a B-17 bomber, the Wellington pilot cut his piston engines, feathered the propellers and opened up the throttle of the W.2B in the tail. The astonished American crew were treated to the incredible sight of a Wellington with two failed engines overtaking them![11] It must have seemed like magic.

As a result of Rover's incompetence, the Meteor didn't fly until 5 March 1943, when the fifth prototype, serial DG206, became the first Meteor to become airborne at RAF Cranwell, piloted by Michael Daunt. The engines weren't Rover engines, though: the Meteor's jet engines had been designed by the brilliant engineer Frank Halford,[12] who had also been responsible for the most powerful piston engine of its day, the 3,055bhp Napier Sabre. Following Whittle's and Rover's ideas, Frank Halford had simplified his centrifugal-flow design with a single-sided compressor and 'straight-through' combustion chambers. Whereas Whittle's chambers doubled back on themselves to reduce turbine entry temperatures, Halford used the new Nimonic superalloy for his turbine blades, which made 'folding' unnecessary (the 'straight through' design was also adopted by the Rover Company for its B.26 engine, later produced in modified form as the Rolls-Royce Derwent).

---

11 Ibid, p. 206

12 Taylor, R., *Boxkite to Jet* (1999). Recommended.

Testing his engine required an enormous air compressor, so a huge air-pumping plant was borrowed at Dartford. This was being used to pressurise the new tunnel under the river Thames, work on which had stopped during the war. The daily bombing raids made this work difficult.

Halford's engine's single-sided compressor revolved at 10,000rpm and compressed incoming air to 40lb/in. This was led to 16 combustion chambers where kerosene was sprayed and ignited. Power developed by the turbine was over 6,000hp, and the exhaust gases travelling at 1,000mph (1,600km/h) provided a thrust of 3,000lb.

The Gloster Meteor prototype was on the tarmac, and the engines were being run up. As the test pilot, Michael Daunt, walked across the front of the aircraft, he was sucked off his feet and dragged into the engine. Fortunately, his head and shoulders became jammed in the inlet, the throttles were closed, and he was pulled out alive. He was lucky to be just bruised and shocked. There have been a number of cases of people who have been sucked into jet engines and who have not survived.

Known initially as the Halford H.1, de Havilland produced this engine as the Goblin, and it eventually had a long life as the Ghost. It would be fitted into the first version of the de Havilland Comet jet airliner.

Whittle's engine had a pressure ratio of around 4 to 1, the Ghost's was only slightly better at 4.6 to 1, so efficiency was poor and fuel consumption was still heavy. Meanwhile, the Metrovick Griffith-designed F.2 jet engine was ready for flight tests in 1943 with a thrust of 2,150lbf and flew as replacement engines on a Gloster Meteor in the November.

Whittle's ideas had finally been adopted by the best engine designers. Too little and too late, at last the British had an effective twin-engined jet fighter.

In Germany, an excited Ernst Heinkel had rushed to his office to announce the triumph of the world's first jet flight. He rang the Air Ministry and was put through to General Ernst Udet, head of the Technical Department. His response was short: 'Everyone is asleep in Berlin,'[13] he said, and hung up.

His country's enemies had cause to be grateful to Udet – eventually. He had been a top flying ace during the First World War, scoring 62 confirmed victories and was the highest scoring German fighter pilot to survive that war, and the second-highest scoring after Manfred von Richthofen, his commander in the Flying Circus. Udet rose to become a squadron commander under Richthofen, and later under Hermann Göring. After the war, he became a stunt pilot and something of an international playboy; he appeared with the notable actress and film director Leni Riefenstahl in three films: *Die weiße Hölle vom Piz Palü* (1929), *Stürme über dem Montblanc* (1930) and *S.O.S. Eisberg* (1933). After this, Ernst Udet went to Hollywood to make films about flying.

Although not particularly interested in politics, he joined the Nazi party in 1933 when Hermann Göring promised to buy him two new US-built Curtiss Hawk II biplanes.

Incidentally, these machines convinced him of the importance of the dive bomber, and against advice from Generalfeldmarschall Wolfram von Richthofen, he pushed for the development of the Ju 87 Stuka dive bomber. Then, late in the aircraft's development, he insisted that the Junkers Ju 88 should also be capable of dive-bombing. The wings were strengthened, dive brakes were added, the fuselage was extended and the number of crewmen was increased to four. No less than 50,000 modifications had to be made, which delayed production and increased the weight from 7 to 12 tons. This resulted in a speed loss of around 80mph (129km/h). In the end, the Ju 88

---

13 Conner, Margaret, *Hans von Ohain: Elegance in Flight*, American Institute for Aeronautics and Astronautics, Inc., Michigan (2001)

proved as vulnerable to RAF fighters as did the Stuka during the Battle of Britain.

Udet was unhappy in his job, hated bureaucracy and turned to alcohol, just as his boss Hermann Göring became addicted to morphine. In April 1941, he led a German delegation inspecting the Soviet aviation industry in accordance with the Molotov-Ribbentrop Pact, just before Hitler's treacherous invasion of Russia two months later, in Operation *Barbarossa*. Udet had informed Göring that the Soviet Air Force and aviation industry were dangerously strong and technically advanced, but he was not allowed to inform Hitler.

Udet became depressed, and on 17 November 1941, he committed suicide by shooting himself in the head while speaking on the phone to his girlfriend, Inge Bleyle. Udet's suicide was kept a secret, and at his funeral, he was lauded as a hero who had died in flight while testing a new weapon. On their way to attend Udet's funeral, the Second World War fighter ace Werner Mölders died in a plane crash in Breslau, and the Luftwaffe executive General der Flieger Helmuth Wilberg died in another plane crash near Dresden. As another German[14] remarked, 'Out of the crooked timber of humanity no straight thing was ever made.'

Ernst Udet's suicide note was written in red pencil. It addressed both his girlfriend and Hermann Göring: 'Ingelein, why have you left me?' and 'Iron One, you are responsible for my death?' Udet's unhappiness had helped to delay the German's jet engine, a delay that would prove crucial.

Ohain's engine did not have a long life. The first He 178 had a fixed undercarriage and was restricted to compressed hydrogen as fuel. The second prototype was better, it burned liquid fuel, but was soon

---

14 Immanuel Kant.

dispatched to a museum in Berlin, where it was eventually destroyed by Allied bombing. This, like the Gloster E.38, was a proof-of-concept aircraft. Heinkel then pressed on with a twin-engined jet fighter design. The significance of Ohain's jet engine had not been lost on others in the German aircraft industry, however. As early as April 1939, the RLM had issued a specification for turbojet aircraft proposals based on a twin-engined single-seat fighter to be constructed of metal. But it wouldn't be powered by Ohain's engines.

The RLM had already foreseen limitations on Ohain's centrifugal turbine design. As we have seen, the large-diameter rotor meant that the engine would have a huge frontal area, resulting in a high drag if fitted to the wings. Straight-through axial-flow engines (like Griffith's design) had smaller diameter compressor turbines, which meant the whole engine had a smaller diameter and thus could be easily fitted to the wings in separate nacelles. Whittle's and Ohain's centrifugal-flow engines meant the use of either single engines built into the fuselage, like the Gloster E.28 and He 178, or engines embedded in the wing roots. They were just too fat to hang below the wings.

There was another disadvantage of the centrifugal-flow type of engine. Both Whittle's original design and Ohain's engine had only one compressor stage. As a result, the ratio between the pressure of air entering and exiting the compressor, the 'pressure ratio', was only around 4.6 to 1 at best.[15] The higher the pressure ratio, the higher the efficiency – a high overall pressure ratio permits a larger nozzle to be fitted on the jet engine. This means that more of the heat energy is converted to jet speed, and energetic efficiency improves. This is reflected in improvements in the engine's specific fuel consumption. Modern subsonic jet engines such as the Rolls-Royce Trent XWB have a pressure ratio of 52:1, while Concorde's engines achieved a ratio of 82:1 while cruising at Mach 2, giving a thermal efficiency with this high-pressure ratio of about 43 per cent, which at the time

---

15 Modern car turbochargers can deliver a pressure ratio of 3.5–4.5:1.

was the highest figure recorded for any normal thermodynamic machine. A piston engine can only manage around 25 per cent.

The axial-flow turbojet could have as many compressor stages as desired, meaning the incoming air could be compressed further and further and thus combustion would be more efficient. Axial turbines could more easily be modified, as stages could be added or subtracted. There was always going to be more potential power available in an axial-flow engine, the Germans realised this early, and this is one reason why their jet engines were more powerful than the Allied centrifugal-flow designs.

And so it was that the RLM encouraged Germany's aero engine manufacturers to begin their own programmes of axial-compressor jet engine development, offering contracts to both Junkers and BMW for an engine capable of 1,520lb static thrust. For the Germans, though, there was going to be one fatal problem: they didn't have enough high-temperature metals.

The planned German jet fighters from the Heinkel and Messerschmitt drawing boards were similar in design as they both carried twin jet engines low under the wings. These were to be axial-compressor designs for the reasons we have seen. The engines would come from Junkers and BMW. At first, there were no restrictions on the expensive materials used by the prototype engines and so they used large quantities of rare metals such as cobalt, nickel and molybdenum.

The Junkers Jumo 004 was the most successful of the two. It was the world's first production turbojet in wartime operation, and it was the world's first successful axial compressor turbojet engine. It was fitted in the Messerschmitt Me 262 fighter and the Arado Ar 234 bomber.

In 1939, with impressive foresight, the RLM contracted the Junkers motor division (known as *Jumo*, a contraction of *Ju*nkers

*Mo*torenbau) to produce a turbojet engine. The Jumo 004 was designed by Dr Anselm Franz to be simpler and quicker to put into production than its BMW competitor, and it was slightly more powerful.[16]

Franz used six combustion chambers, or 'flame cans', instead of a single, more efficient annular chamber. He built an eight-stage axial-flow compressor, with the six axial combustion chambers made from cheap sheet steel and a one-stage exhaust turbine with hollow blades. Air bled from the compressor passed through them to provide cooling. He also went straight into the prototype stage rather than experimenting with development engines as Whittle had done. The Germans had a far greater sense of urgency than the British about their jet programme.

Perhaps surprisingly, their jet engines turned out to be simpler and cheaper to make than piston engines, employing lower-skilled workers using simpler tools. The sheet metal combustion cans and hollow turbine blades could be made on the same tooling as that used for making car body panels and engine parts.

Another advantage of the new engine was that it could run on a variety of fuels. Its standard fuel was a liquid synthesised from brown coal: J2. It could also run on paraffin (kerosene) or diesel. It *could* run on aviation gasoline, but that was considered too precious to use in any engine that didn't need it. Piston aero engines, of course, demanded the best high-octane petrol (gasoline). On the other hand, jets are inherently thirstier than petrol engines. Owing to the Allied blockade, the Germans suffered from a scarcity of fuel throughout the war.

Dr Franz's first prototype of the world's first production turbojet was first bench tested in October 1940. Like Whittle, he ran into

---

16 Boyne, Walter, *The Jet Age: Forty Years of Jet Aviation*, Smithsonian, Washington, DC (1979), pp. 25–46. The Jumo 004 B-1 produced 1,980lb thrust, the BMW 003 A-1 gave 1,760lb (800kg), whereas the British Power Jets W.2/500 produced 1,600lb thrust. The American General Electric I-16 also gave 1,600lb.

problems, in this case vibration from the aluminium stator blades. A switch to stronger steel blades solved that problem.

Meanwhile, test flights began of the world's first jet fighter, the prototype Messerschmitt Me 262. To test the airframe and controls, on 18 April 1941 it flew with just a conventional piston Junkers Jumo 210 engine mounted in the nose and driving a propeller. Later, two prototype BMW 003 turbojets were hung under the wings for in-flight testing but, wisely, the piston engine was retained. Not only one but both BMW 003s failed in flight, flaming out, and so the test pilot had to return to base on just the piston engine.

Like Rover's, the BMW jet engines were proving problematic.

On 15 March 1942, the Junkers Jumo 004A jet engine was flown in a Messerschmitt Bf 110 and tested in flight, and on 18 July 1942, the Messerschmitt Me 262 flew for the first time under jet power from its 004 engines. This was almost nine months ahead of the Gloster Meteor's first flight on 5 March 1943.

It would be over a year until the Meteor saw action, when on 27 July 1944, the British jet was high over Kent, chasing a new terror weapon, the V-1 Doodlebug jet-engined flying bombs. Two V-1s were shot down on 4 August, and so the Allies were able to study this new weapon of war.

The V-1 *Vergeltungswaffe 1* or 'Vengeance Weapon 1' was an early form of cruise missile or flying bomb, and it mounted a valved pulsejet engine. This used mechanical reed valves rather like a Venetian blind across the air inlet. Air entered the open reeds; petrol fuel was ignited in a combustion chamber and the reed valves then slammed shut. The hot gases of combustion were therefore obliged to exit the back of the engine through the tailpipe, producing thrust, and this allowed a fresh charge of air to enter through the intake. The inertia of the escaping exhaust created a partial vacuum for milliseconds after each detonation. This drew in additional air between the combustion pulses, which cycled at around 42 times a second, giving rise to a loud buzzing exhaust note rather like a motorcycle engine. The distant buzzing as it approached led to the V-1 being dubbed the 'Buzzbomb' or 'Doodlebug'. Hitler called it the 'Maybug'.

The result was a crude but effective jet engine, working something like a piston engine with sequential combustion, but with far fewer parts. The pressure ratio was only around 1.2:1, meaning that this engine could never be efficient. The rest of the V-1 was as might be expected: a sheet-steel fuselage, wooden wings and a guidance system composed of two gyroscopes controlling yaw and pitch. A magnetic compass-controlled azimuth. Altitude was maintained by a barometric cell. The warhead was nearly a ton of Amatol. When a distance-run device triggered, the V-1 went into a steep dive, the engine cut out and the British citizens knew it was time to run for cover.

The V-1s flew at around 400mph (640km/h), at between 2,000ft and 3,000ft. Over 1,000 V-1s were destroyed by aircraft, but the only Allied aircraft that could catch and overtake them were a handful of Napier Sabre-powered Hawker Tempests (638 victories), Mosquitos (623) and the jet-engined Gloster Meteor (13). The Merlin-engined Spitfires had to dive before they could reach the same speed as the V-1. Griffon-engined Spitfires did rather better (303), and so did the P-51 Mustangs (232). Unfortunately, the new Meteors suffered from jamming guns and accounted for only 13 of the V-1s.

Pilot Tony Blackman, who trained on piston-engined aircraft, remembered how different the Meteor was: 'When you got into a Meteor the first thing that struck you was the fantastic view all the way round. You could see what you were doing.'

RAF pilot Norman Tebbit also remembered the excitement of taking off in a Meteor:

There was nothing in front of you. Whereas flying a conventional single piston engined aircraft; the Hurricane or the Spitfire, there's a huge great engine in front of you... it was so exciting. Get airborne, up with the wheels, hold it low until you were about 380 knots, pull it up and she would go up, well we thought then, like a rocket.

After the end of the V1 campaign, RAF pilots were forbidden to fly the Meteor over enemy-held territory in case an aircraft was shot down and salvaged by the Germans (or indeed the Russians). But after a great deal of effort and expense, the British had squandered Whittle's technical lead.

A word about knots, and why jet pilots use them. A knot is a speed of one nautical mile per hour. The nautical mile is based on the circumference of the Earth: imagine that the equator is a circle divided into 360° (like a compass). Each degree is then split into 60 equal parts called minutes. The length of each of these minutes is equal to one nautical mile. The Englishman Robert Hues invented the measurement, proposing in 1594 that the distance along a great circle was 60 miles (97km) per degree, that is one nautical mile per arcminute. Calculating distances as fractions of the Earth's surface is more convenient for navigators, both on ships and in aircraft.

Speed was difficult for ancient mariners to measure. At first, a system akin to Poohsticks was used: a chip of wood was dropped off the bow and the time it took to reach the stern was measured. A 50ft boat travelling at one nautical mile per hour would pass the floating chip in roughly 30 seconds. By the time of Nelson, this had been improved: a long line with knots tied at 48ft (15m) intervals was attached with a pegged bridle to a triangular panel of wood called a log, which was heaved overboard. The log was weighted on one edge, so it floated upright more or less stationary. The line therefore ran out at the same speed as the ship. A 28-second sandglass was turned, and the sailor counted the number of knots that passed through his fingers during the 28-second interval. At the end of the allotted time, he nipped the line with his fingers, and the number of knots running out in the 28 seconds represented the ship's speed. A yank on the line pulled out a wooden peg on the bridle, which enabled the log to be pulled in flat, bouncing across the waves. At speed in a storm, this was a hazardous operation. An old salt's tale tells of a young sailor on a tea clipper in a storm. He was holding the line-reel in his hands and was snatched over the side when the ship's speed exceeded the length of the line. And the speed? 'Sixteen knots and the boy, sir.'

After its successful flight, the Jumo jet engine was ordered into production by the RLM. So far, so good, but this is where the German jet programme ran into problems.

Those first Jumo 004A engines had been built with no regard to expense, with liberal use of scarce metals such as nickel, molybdenum and cobalt. The Allied blockade had been effective in reducing Germany's access to these strategic materials, and Dr Franz knew that, once his engine had been proved, he had to find some way to make it more cheaply. His second, 004B, engine was redesigned to incorporate as few of these precious materials as possible. Many of the metal parts exposed to high temperatures were changed to mild steel, such as the combustion cans, which were made of mild steel sprayed with aluminium (you'll still see this finish on modern motor car exhausts). But the exhaust turbine blades, which had to endure the highest temperatures and rotational speeds, still had to be fabricated from superalloys.

These exotic metals would become central to the story of the turbojet. Superalloys are the nickel-, cobalt- and iron-based alloys used in the hottest, most demanding components in gas turbines, working at over 1,000°C. Not only must they resist melting, but they must also resist creep – the tendency of a solid material to deform permanently under the influence of steady mechanical stress. You'll see creep in a glacier as it flows down a mountain. Creep is more severe in materials that are subjected to heat for long periods and generally it increases as they near their melting points. If a turbine blade creeps, stretches a fraction of an inch and touches the casing, it will destroy the engine and could bring down an aircraft.

Dr Franz used an alloy named Cromadur, developed by Krupps and made of 12 per cent chromium, 18 per cent manganese and 70 per cent iron. This was folded and welded, and the hollow turbine blades were cooled by pressurised air bled from the compressor. As we have seen, Frank Whittle used a superalloy called Nimonic 80 that had been developed by the Wiggin company, which started from the pre-war Nichrome formula (80 per cent nickel, 20 per cent chrome) made for electric-fire elements, a material that resisted oxidation at bright cherry red temperature but, of course, with no stress. You can see this stuff glowing red-hot if you peer into your electric toaster in the morning. By adding small amounts of titanium, aluminium, iron, carbon and then heat-treating, the creep resistance at useable stress was increased.

This first production German turbojet engine had a casing made in two halves of cast magnesium, a metal 33 per cent lighter than aluminium but rather prone to catching fire. The casing halves had half-sections of stator blades bolted to them, rather like slices of toast sticking out of a toast rack. When closed together, the two halves interleaved the rings of stator blades with the rotating compressor blades.

The new jet engine was surprisingly cheap to make, once the scarce metals were stripped out. A Junkers Jumo 213 piston engine, similar to the Rolls-Royce Merlin took around 1,400 hours' labour to build and cost 35,000 Reichsmarks, whereas the Jumo 004B jet engine took only 375 hours and cost only 10,000 Reichsmarks. Moreover, the jets used lower-skilled labour, a vital factor towards the end of the war when slave labour was used.

Late in 1943, this production 004B engine started to suffer turbine blade failures which perplexed the Junkers team. They had to call up a turbocharger specialist, Max Bentele, who found that the failures were being caused by sympathetic vibration: one of the blade's natural frequencies was in the engine running range. This was similar to the failures that had bedevilled the Mercedes Benz six-cylinder piston engine crankshafts 40 years before. Bentele solved the problem by raising the frequency of the blades: shortening them by 1mm and by reducing the operating speed of the engine from 9,000 to 8,700rpm.

Dr Franz's lower quality 004B engine only had a service life of ten to 25 hours. In comparison, Whittle's W.2/700 turbojet engine had an operational life span of 125 hours. Whittle assessed the two engines after the war and concluded: 'It was in the quality of high temperature materials that the difference between German and British engines was most marked.' This is not quite true, as the Germans had clearly mastered the superior concept of the axial-flow jet engine. But the lack of superalloys was a huge obstacle to the German turbojet.

That problem became much worse towards the end of the war, as nickel and chromium supplies petered out. Many of the newly unpacked engines at the airfields didn't even make it through the onboard testing procedure before failing and needing to be replaced.

The 004B also suffered from a slow throttle response, a problem suffered by all early turbojets. An unskilled pilot could wreck an engine by opening the throttle too sharply, allowing too much fuel into the engine which would overheat the turbine blades.

Amusingly, the Jumo 004 had a D-shaped handle on the front of the intake, for all the world like the ripcord starter on a lawnmower engine. This would, in fact, start a small 10hp two-stroke piston engine, which, in turn, would revolve the turbine shaft and start the jet engine: and this was the world's first auxiliary power unit (APU) for a jet. Two small tanks of petrol/oil fuel were neatly installed in the upper portion of the annular intake's sheet metal housing. This little gem of an engine had an ultra-short stroke and sideways sparkplugs to fit in the jet engine's hub. Neat.

Its designer, Norbert Riedel, went on after the war to make scooters. How are the mighty fallen! Just as Heinkel made bubble cars after jet bombers, so Messerschmitt made three-wheeled cars[17] after fighters, and Porsche made sports cars after tanks. Riedel died in an avalanche in 1963.

Full production of the engine and, therefore, the Me 262 was delayed until 1944, by which time Allied bombers were over German cities night after night. They had to get this new fighter into the air.

The German jet aircraft development programme had been dogged throughout by political interference. Even as early as February 1940, the German High Command had decreed that any experimental project not in operation between six months and a year should be scrapped: they did not expect the war to last that long. The following year, in a *Führerprotokoll* (Führer directive), Hitler repeated that all experimental projects that could not be completed within six months were to be terminated. But Göring wisely created an exception for the jet programme, and both Heinkel and Messerschmitt pressed on with their jet fighters. After the prototypes had been successfully completed, the RLM allowed them to continue.

By the time the Messerschmitt 262 prototypes were starting to show their astounding speed, Hitler interfered again. The aircraft had been designed in two versions: a fighter named Schwalbe (Swallow) and a fighter-bomber called Sturmvogel (Storm Bird). Fearing an Allied

---

17 True to form, the Messerschmitt factory prepared a KR200 Kabinenroller (cabin scooter) to break the 24-hour speed record for three-wheeled vehicles under 250cc (15.3 cu in). The record car was run on 29–30 August 1955 at the Hockenheimring and set a new 24-hour speed record at 64mph (103km/h).

invasion of France, Hitler decreed in 1943 that all efforts should be concentrated on the fast bomber, Sturmvogel, version.

Although 1,400 Me 262s were produced, there was only a maximum of 200 in operation at any time. This was due to endless technical problems with the new airframe and the completely novel power plant.

Production was ordered to be stepped up, and because of Allied bombing, dispersed factories and underground factories had to be built. Wings were made in Germany's first autobahn tunnel. And as we have seen, slave labour was used to make the Me 262 – at the Gusen II concentration camp, workers produced 450 finished Me 262 fuselages a month. The treatment of workers was brutal, and the typical life expectancy was six months. Subtle sabotage of airframes by disaffected workers was a constant danger. An estimated 35,000 to 50,000 people died while forced to build the Me 262.

In his memoirs, Albert Speer, the Minister of Armaments and War Production, maintained that Hitler had blocked mass production of the Me 262, then changed his mind and lifted the restriction in early 1944 when the course of the war started going badly for Germany. According to Speer, Hitler wanted the fast bomber to be used for revenge attacks rather than be used as a fighter to fend off the Allied bombers, which were pummelling the German homeland. But it now appears that the engine's vibration problems delayed the appearance of the Me 262 as much as Hitler's own slower-frequency vacillation.

The Messerschmitt 262 was slowly introduced to training squadrons during 1944 and pilots learned the limitations of the technology. The first characteristic they discovered was poor performance of the Jumo 004 engine at low speeds, particularly at take-off and on landing. The high-altitude performance got better and better during the rest of the war, but the low-speed characteristics remained poor.

As soon as the Allied piston-engined fighters realised this, they loitered around the airfields and picked off the Me 262s when they were at their most vulnerable.

'The Messerschmitt Me 262's most dangerous opponent was the British Hawker Tempest,' reported Hubert Lange, a Me 262 pilot, 'extremely fast at low altitudes, highly manoeuvrable, and heavily armed'.[18]

The Germans responded by providing the Me 262, the fastest fighter in the world, with a piston-engined fighter escort. And so the Luftwaffe was forced to use piston-engined fighters to defend jet-engined fighters!

In early 1945, Charles E. 'Chuck' Yeager of the 357th Fighter Group, an American fighter pilot, was stooging around an Me 262 airfield in his P-51 Merlin-engined Mustang, *Glamorous Glennis*:

> I spotted a lone 262 approaching the field from the south at 500 feet. He was going very slow, about 200mph. I split-essed on him[19], and was going around 500mph. Flak started coming up very thick and accurate. I fired a single short burst from around 400 yards and got hits on his wings. I had to break straight up, and looking back saw the enemy aircraft crash-land about 400 yards short of the field. A wing flew off outside the right jet unit. The plane did not burn.

After the war, Chuck Yeager would go on to become the first pilot in history confirmed to have exceeded the speed of sound in level flight.

---

18 http://www.hawkertempest.se

19 The Split S involves half-rolling the aircraft inverted and then executing a descending half-loop, resulting in level flight in the opposite direction at a lower altitude.

Once it gathered speed though, the new Me 262 jet fighter suddenly made piston-engined aircraft like Yeager's Mustang look as if they were going backwards. It could reach 560mph (900km/h), around 100mph (160km/h) faster than the Allied fighters, and could climb to 38,000ft at a rate of 3,940ft per minute with a range of 650 miles (1,050km). It was armed with four 30mm MK-108 cannons, which could fire more than 650 rounds per minute, and it could carry 1,000lb (450kg) of bombs.

When the Messerschmitt 262 did appear in the daytime skies over Germany, it caused consternation among the huge box formations of USAAF B-17 Flying Fortress bombers intent on bombing the Fatherland.

The idea behind the 'combat box' was that the B-17s would be protected by interlaced .50in calibre machine-gun fire from their gunners, thereby allowing the Flying Fortresses to defend themselves against enemy aircraft during daylight raids. A serious disadvantage was that piston-engined enemy fighters attacked the highest and lowest first, and to be placed at the end of the bottom level as 'tail-end Charlie' was a death sentence. The new jet plane added two more deadly stratagems.

General Arnold said, 'The jet propelled airplane has one idea and mission in life and that is to get at the bombers, and he is going by our fighters so fast that they will barely see him, much less throw out a sky hook and slow him up.'

The new jet was almost too fast to attack the bombers, as head-on speeds were over 700mph (1130km/h) – far too fast for accurate shooting. And so, the pilots evolved a special 'roller-coaster' technique: they approached the formation of bombers and escorting fighters from astern in a shallow dive that took them straight past the fighters who could not hope to catch up. When below and behind the bombers, they pulled up sharply to wash off speed; they then machine-gunned their targets.

'This was a Blitzkrieg aircraft. You whack in at your bomber. It was never meant to be a dogfighter; it was meant to be a destroyer of bombers...', reported Allied test pilot Captain Eric Brown after testing a captured Me 262.

> The great problem with it was it didn't have dive brakes. For example, if you want to fight and destroy a B-17, you come in on a dive. The 30mm cannon weren't so accurate beyond 600 yards So, you normally came in at 600 yards and would open fire on your B-17. And your closing speed was still high and since you had to break away at 200 yards to avoid a collision, you only had two seconds firing time. Now, in two seconds, you can't sight. You can fire randomly and hope for the best. If you want to sight and fire, you need to double that time to four seconds. And with dive brakes, you could have done that.[20]

Ironically, Udet's demand for dive brakes on other aircraft would have been of more use on the Me 262. But he had killed himself by then.[21]

When the German R4M rockets became available, the jet fighter was even more lethal. They would now approach the bomber formation from the side and fire 24 rockets from under their wings while they were still out of range of the defending machine-gun fire. Luckily for the young American aircrews, only a few Me 262s were fitted with rockets and these attacks came late in the war. The British crews of the night-bombing Avro Lancasters didn't have to face so many attacks from the new terror weapon until a night fighter version of the Me 262 was developed.

---

20 Thompson, Steve J. with Smith, Peter C., *Air Combat Manoeuvres*, Hersham, Ian Allan Publishing (2008)

21 One wonders if slowing the bombers down to near-stalling speed while under attack might have made them harder to hit with a fast-moving jet plane.

Rockets, of course, carry their own oxidising agents in their propellant, which is why they can work in the vacuum of space. Jet engines require oxygen from the air.

In Britain, some of the Gloster Meteors, which were similar in appearance and performance to the Me 262, were taken off their V-1 flying-bomb-chasing duties to help their American allies. Frank Whittle tells us they were assigned to tactical trials in British airspace to help work out a way to defend the Flying Fortresses against the German jet fighter:

> Results of these trials must have proved very depressing to the Americans with their existing aircraft, as it would appear that the Meteors could sail in as they pleased, each 'destroying' two or three Forts and pull away without the escorting fighters (even Mustangs) being able to do very much about it.[22]

The last great air battle over Germany came on 18 March 1945, with a raid on Berlin including over 1,220 Allied bombers and scores of Merlin-engined P-51 Mustang fighters up against heavy German flak and Me-262 jet fighters using R4M rockets.

Oberleutnant Günther Wegmann was the commander of Jagdgeschwader 7's 9th Squadron of Me 262 jets. He led his squadron in a loose formation towards the incoming bombers. Three of his jets fired their R4M rockets into a box formation of 60 Boeing B-17 Flying Fortress bombers from a distance of 3,000ft. The scores of rockets created devastation, with bits of aircraft, smoke and flame erupting from the formation of bombers.

'The rockets gave us extra punch', said Me 262 pilot Leutnant Klaus Neumann. 'Fire the rockets, do the damage, weaken the tight formation integrity of the bombers, and then pick off the crippled stragglers. It was like being a god in a way.'

---

22 Whittle, Frank, *Jet*, Muller, London (1953)

Chapter 4

The Jet that Shook the World

In the years following the Second World War, the British aircraft industry flourished. The skies were filled with jet fighters and jet bombers, and the aircraft and the men who flew them were the stars of this new age.

At 0241hrs on the morning of 7 May 1945 at Supreme Headquarters Allied Expeditionary Force (SHAEF) headquarters in Reims, France, General Alfred Jodl had signed an unconditional surrender document for all German forces to the Allies. General Walter Smith signed on behalf of the Supreme Commander of the Allied Expeditionary Force, and General Ivan Susloparov on behalf of the Soviet High Command. This marked the end of the Second World War in Europe.

However, Stalin was furious, stating that Susloparov had no authority to sign, and he demanded a second surrender. A slightly modified document was therefore signed on 8 May 1945 in Karlshorst, Berlin, at 2120hrs local time. This Soviet intransigence was to be symbolic of the beginning of the freezing of relations between the former Allies.

The coming Cold War was to be dominated by jet aircraft.

In occupied Germany, the former Allies raced to loot the best Nazi technology. US Air Disarmament squadrons worked rapidly to identify, secure and carry off Luftwaffe items. Much of this work was secretive, with the Americans driving trucks to the sites of German V-2 rockets and collecting 100 V-2s before the Soviets got there, another action that enraged Stalin. Some were more overt, such as Operation *Lusty*, during which serviceable Me 262s were rapidly 'bagged' by the USAAF.

The Americans, Soviets and British all wanted to evaluate the technology, particularly the turbojet engines. Russian soldiers dismantled entire German aircraft factories, even to the extent of numbering each brick and shipping them back east. They also captured German scientists, engineers and technicians. British troops entered the Focke-Wulf facility in Bad Eilsen and confiscated plans, models and wind-tunnel data on the planned successor to the Messerschmitt Me 262, the TA-183 jet fighter. These they shared with the Americans. Meanwhile, Soviet forces searched the Berlin office of the German Air Ministry and found a complete set of TA-183 blueprints and a mass of wing research. The Soviets were determined that Western technology was going to help the Soviet Union to catch up in jet science.

Back in Britain, comparison tests were made. The German Me 262 was found to be faster than the British Gloster Meteor jet fighter, and it had better visibility to the sides and rear. It was a more stable gun platform than the Meteor, which had poorer directional stability at speed, with a 'snaking' characteristic. On the other hand, the Meteor had a longer range and much more reliable engines: the Me 262 needed no less than four changes of engine during four hours of test flights. The British jet engine technology was demonstrably better, and that was something the British were going to be able to sell.

Nations around the world celebrated the Allied victory and began to repair the damage of wartime austerity. There was a huge feeling of relief and exuberance in Britain. Within two months, a Labour government was in power after a surprise landslide victory; Clement Atlee had promised to provide jobs for returning servicemen and to rebuild the British economy. This government introduced the National Health Service,[1] implemented the National Insurance Act and promised to provide welfare 'from cradle to grave'. Ex-service personnel voted Atlee in.

During the war, there had been a million men and women working for over 30 aircraft companies, and the RAF was basking in the nation's gratitude. All that wartime investment, expertise and goodwill could be put to the new challenge: Export or Die. After all, Whittle had invented the jet engine, and this was going to propel Britain into an exciting new age. In 1945, the Gloster Meteor was the state of the art in jet aviation, as the Messerschmitt Me 262 was now grounded after Germany's defeat. As a result, the Meteor's Rolls-Royce engines were going to have an international market.

Rolls-Royce, with its huge resources, had pushed on with development of the turbojet at impressive speed. Based upon designs produced by Power Jets, Rolls-Royce produced more advanced turbojet engines with more power. This is what Frank Whittle could have done with all along. Continuing the river-named theme, the Derwent[2] engine succeeded the Welland. The Derwent Mark I entered production with 2,000lbf (8.9kN) of thrust and was fitted to most early Gloster Meteors.

A Mark II Derwent was modified with a smaller impeller, with the turbine unchanged, and with a shaft running forward from the impeller to a reduction gearbox driving a five-bladed propeller. This was named the Rolls-Royce RB.50 Trent and was the first turboprop to fly, in a test bed Meteor. Propellers improve take-off and climb performance, and so turboprop engines are better for aircraft operating on short airstrips. Propellers are most efficient at slow and medium speeds and become less efficient as the speed of the aircraft

---

1 The NHS, pushed through by ex-miner Nye Bevan, opened its doors on 5 July 1948.

2 The River Derwent flows through the city of Derby, the site of the Rolls-Royce headquarters. The Derwent valley was the cradle of the Industrial Revolution, and the Derwent provided power for the first cotton mills.

increases. Turboprop engines also tend to be more fuel-efficient than turbojet engines.

The Gloster Meteor was used as a development mule for other early turbojet designs – one being used for reheat or afterburner testing, where neat fuel is injected into the jet pipe, providing a massive increase in thrust for take-off (this was another Whittle innovation). Another Meteor was fitted with the Rolls-Royce Avon, a new axial-flow engine. Other companies involved in the development of the Meteor's engines, Armstrong Siddeley, de Havilland, Bristol and Metropolitan-Vickers also speedily developed their own turbojet engines.

Now everyone wanted a slice of Whittle's pie.

The Rolls-Royce Derwent jet engine had one unusual application: four of them were used to power the former Clyde paddle steamer *Lucy Ashton*. This 1888 ferry had its steam machinery removed and replaced by the Derwents in 1950–51. They provided a thrust of 3600lbf each at 14,500rpm. The idea behind this was to research the amount of drag on a ship's hull in real-life conditions, and jets were preferred to underwater propellers or paddles as these would have created a disturbance in the water, and the force exerted by them was harder to measure, as it would have involved strain gauges on thrust bearings buried in the bilges. The *Lucy Ashton*, which was named after the opera singer sister of Lord Henry Ashton, roared unmusically up and down a sea loch conducting trials at over 15 knots and infuriating the local residents. Marion Gillies said, 'She sped up and down the Gareloch for two years, and the racket was unbelievable.'[3]

3 Another unusual use was for a Derwent Mark 8 from a Gloster Meteor, which was used in the 180mph (290km/h) jet-propelled car Thrust 1, which was built by Richard Noble in 1977. On its second run the car rolled at 140mph (220km/h) and destroyed the Derwent.

Stanley Hooker, the genius behind the Rolls-Royce centrifugal supercharger, had been given the job of meeting a 1944 Air Ministry demand for an engine of 4,200lb-force of thrust (lbf). After visiting the United States, he saw that General Electric already had two jet engines in development: one centrifugal like Whittle's and one axial. They were already running and were expected to develop 4,000lbf. On his return to Rolls-Royce, he decided to aim for 5,000lbf (22.24kN) of thrust and designed a new centrifugal jet engine with a double-sided impeller, nine improved combustion chambers and other tweaks that gave the desired thrust.

The early days of any technology often see huge gains, just as they did with the piston aero engine, which went from the 14hp of the Wright brothers' Flyer to the 2,200hp of the Rolls-Royce 'R' Schneider Trophy winners in just 25 years. Turbojet development was no different, and the sky seemed the limit.

The new engine was named the Rolls-Royce Nene,[4] and it was licensed to Pratt & Whitney and fitted to the American aircraft company Grumman's first jet plane, the F9F Panther. But the Rolls-Royce Nene had a curious history: British technology was inadvertently going to help the Soviet Union to field one of the world's best jet fighters.

The Soviets were developing a new fighter, the Mikoyan-Gurevich MiG-15, and they needed a more powerful and reliable jet engine. The looted Jumo 004s were too unreliable and had insufficient thrust for the performance required. They tried securing engines from Britain and the United States without much luck, until in 1946 the Atlee government discreetly signalled that it would like to improve post-war relations with the Soviet government.

---

4 Another East Midlands river, the Nene forms the boundary between Cambridgeshire and Norfolk. Henry Royce was born on its banks at Alwalton and his ashes are buried there. The pronunciation of the river's name is *Nen* as in *hen* down to Thrapston, where it changes to *Nene* as in *queen*. Oddly, the point of this transition has moved steadily upstream over the years.

Mikhail Khrunichev, the Soviet aviation minister, and his aircraft designer, Alexander Yakovlev, suggested to Stalin that the USSR should buy the Nene engines from Rolls-Royce for the purpose of copying them. Stalin is alleged to have replied, 'What fool would sell us his secrets?'

Sir Stafford Cripps, who had Marxist sympathies, was the then head of the Board of Trade and invited Khrunichev and Yakovlev as part of a Soviet trade delegation to tour British aerospace factories. The presence in that delegation of Artem Mikoyan himself (the 'Mi' in MiG) should have alerted the British to what was going on. With the currency constraints then in force, the Russians offered to buy the Derwent and the Nene for gold bullion, and Cripps quickly arranged to have the Nene taken off the secret list. Rolls-Royce was given permission in September 1946 to sell ten Nene engines to the USSR, and a further 15 in March 1947. The price was fixed under a commercial contract. This directed Rolls-Royce to sell Nenes to the Soviet Union on the condition that they were not to be used for military use.

However, the Soviets secretly reverse engineered the British engine, taking the engines apart and copying each individual part. Later, this subterfuge was discovered by Whitney Straight, then deputy chairman of Rolls-Royce, during a visit to Beijing. He saw that the Rolls-Royce engine had been copied without licence. The Soviets had built the Klimov RD-45 jet engine and a larger version, the Klimov VK-1, which soon appeared in the MiG-15 jet fighter. This had a maximum thrust of 5,955lbf (26.5kN).

Why did the British government officials agree to such a sale? Were they naïve? Greedy? Or were their political sympathies too closely in tune with Joseph Stalin? He was, after all, probably the most monstrous dictator ever, his regime being responsible for 9 million deaths, at least 6 million of which were deliberate killings.

There is more to this story than meets the eye. Although the Nene was the most powerful jet engine in the world, there were few orders for it. The government had consulted General Hastings Ismay, Chief of the Defence Staff, and he explained that the RAF had no interest in the centrifugal Nene engine, because it knew there was something better coming: axial engines. The secret Rolls-Royce Avon was reputed to be capable of 6,500lbf (29kN) thrust, with much more to come.

Atlee was desperate to make good his election promises to the British people and find markets for British exports. After all, hadn't Churchill himself promised to sell British steel to the Soviets before he left office? And Britain was nearly bankrupt.

Rolls-Royce and the government would have far preferred to sell their technology on the basis of a licensing production agreement, as they had with the Americans. The Russians, on the other hand, only wanted a small number of engines. And they were offering gold. The British didn't consider the Russians to be capable of reverse engineering anything as complex as a state-of-the-art jet turbine, particularly in view of the metallurgy of the superalloys involved, and so the deal was done. In this the British underestimated the Soviet ability.

But there might have been an ulterior motive. Might having an intimate knowledge of the engines of Soviet jet aircraft have been worth being involved in their development?[5]

In the end, if Atlee's administration is judged on its record, it is fair to point out that it had the defence of the nation at its heart, if you consider the enormous cost and effort put into the development of nuclear weapons and the jet-engined V bombers to deliver them.

---

5 For the arguments, see Bryen, Stephen D., *Technology Security and National Power: Winners and Losers*, Routledge, New York (2017)

War is a powerful engine of innovation. The 1950 Korean War was the first in which jet aircraft on both sides played the deciding role in aerial combat. It is a war that has somehow become forgotten, even though 3 million died, including 33,000 US military combatants. The United States and Russia fought a proxy war over Korea, and a new superpower, China, took its first baby steps. All that remains in the public imagination is the *M*A*S*H* film and TV series, and dimly at that.[6] Most viewers seem to think *M*A*S*H* had something to do with Vietnam.

At the end of the Second World War, the United States and the Soviet Union had liberated the Korean peninsula from the defeated Japanese. Just as in Germany, the two competing allies had divided the country into two zones of occupation, in this case at the 38th parallel of latitude. A communist state was established in the north and an anti-communist state in the south.

In Jonathan Swift's satire, *Gulliver's Travels*, the two countries of Lilliput and Blefuscu are at loggerheads over which end of an egg should be cracked at breakfast. This led to a murderous war between the Big-endians (Blefuscu) and the Little-endians (Lilliput). Swift's satire was no less relevant in 1940s Korea: in the north, a communist Emperor, Kim Il-sung, had set up a dynasty, which still endures, and in the south, there was an anti-communist state led by Syngman Rhee. Both, of course, claimed to be the sole legitimate leaders of Korea.

Just as the enormous Gulliver assisted the tiny Lilliputians in their invasion, so too did the communist Soviet Union and China aid North Korea in its invasion of South Korea on 25 June 1950. The United Nations (UN) Security Council, condemning this as illegal, authorised the dispatch of forces to Korea. Twenty-one countries

---

6 The film's director, Robert Altman, stipulated that the theme song 'Suicide is Painless' had to be the stupidest song ever written. Attempting to write it himself, he despaired and gave the job to his 14-year-old son, Michael, who wrote the lyrics in five minutes. Atman earned US$70,000 from the film. His son earned US$1m from the song.

of the UN eventually contributed to the UN force, with the United States providing around 90 per cent of the military personnel.

At the beginning of the war, the aircraft facing each other were Second World War piston-engined fighters. On the UN side were the Rolls-Royce Merlin-engined P-51 Mustang, the Pratt & Whitney Double Wasp-engined F4U Corsair and the Napier Sabre-engined Hawker Sea Fury. And at first on the North Korean side were the Soviet Yakovlev Yak-9 and Lavochkin La-9s.

The Yak-9 did not do well: one page of the aircraft's operating manual was accidentally omitted from the translation from Russian. Before starting the Yak-9, it was necessary to hand-crank a small cockpit-mounted oil pump 25 times to provide initial lubrication to the Klimov M-105PF V12 engine. When North Korean pilots missed this crucial pre-lubrication, the engine was prone to seize just after take-off, causing several fatalities. Even when they were aloft, neither aircraft was powerful enough to be competitive. Neither the Yak-9 nor the Lavochkin La-9 could stand up to the UN piston fighters.

Then suddenly the tables turned. On 30 November 1950, a US Air Force (USAF) B-29 Superfortress was attacking an air base in North Korea when it received hits from a fighter that overtook the bomber so fast it could not be identified. Straight-wing Lockheed F-80 jets escorting the bomber attempted to chase the mysterious fighter, but it disappeared. The Chinese had just introduced the new Soviet-built MiG-15 fighters with their Klimov RD-45 jet engine copied from the Rolls-Royce Nene.

These also featured a new swept-back wing, which made the small fighter extremely fast. It was able to reach the American B-29 bombers and shoot them down. On one disastrous day in October 1951, MiGs shot down six out of nine Superfortresses. From that November, B-29s stayed on the ground during the day and bombing missions were flown only at night.

'Oh, it was a wonderful airplane', said retired USAF Lieutenant General Charles 'Chick' Cleveland, remembering his first encounter with a MiG-15. 'You have to remember that the little MiG-15 in Korea was successful doing what all the Focke-Wulfs and Messerschmitts of World War II were never able to do: drive the United States bomber force right out the sky.'

The speed of the MiG-15 shocked its UN opponents. The British Chief of the Air Staff said, 'Not only is it faster than anything we are building today, but it is already being produced in very large numbers... The Russians, therefore, have achieved a four-year lead over British development in respect of the vitally important interceptor fighter.'

Swept wings have the effect of delaying the shock waves and the consequent aerodynamic drag rise caused by compressibility near the speed of sound, hugely improving performance at around 600 to 700mph (970 to 1,130km/h). So, they are useful at transonic speeds, which are the speeds that airliners tend to fly at (they are not compulsory for supersonic flight: Chuck Yeager's Bell X-1 had straight wings).

Although the Second World War Messerschmitt Me 262 had slightly swept wings, they were there because the engines had turned out heavier than expected, and the 18.5 per cent angle corrected the centre of gravity. Swept-back wings were nothing new, as the British aircraft designer John Dunne had built a tail-less biplane with heavily angled-back wings to provide longitudinal stability back in 1913.[7]

The Soviets and Americans would have known that designer Adolf Busemann had suggested fitting a 35° swept wing to the Me 262, the same wing-sweep angle later used on both the American F-86 Sabre and Soviet MiG-15 fighter jets. Though his suggestion wasn't accepted for the production Me 262, he pursued the idea and designed

---

7 Dunne's swept wing was even cleverer than it looked, as he also incorporated wash-out in the wings: the angle of incidence was greater at the wing roots and decreased along the wing, becoming least at the wing tip. He did this to ensure that at stall speed the wing root stalled before the wing tips, thereby maintaining aileron control. This feature was used on the Supermarine Spitfire.

high-speed (Hochgeschwindigkeit) versions, the HG II and HG III, which had a 35° and 45° wing sweep, respectively. And the planned successor to the Me-262, the Focke-Wulf Ta 183 Huckebein, would have had 40° swept wings and a high tail. These designs could have been considerably faster than the original Me 262, as the MiG-15 demonstrated.[8]

The UN's first-generation straight-wing jets of Western air forces were the Lockheed P-80 Shooting Star (the first US jet fighter), the Nene-engined Grumman F9F Panther and the British Gloster Meteor. They were all out-matched by the new swept-wing MiG-15. In this new jet-fighter conflict, much the same design of Whittle-based centrifugal jet-engine was fielded on both sides, but the swept-wing configuration had a clear advantage.

United States hastily introduced the North American F-86 Sabre, another single-engine jet fighter with wings swept to 35°, stubby fuselage and T-tails. The two aircraft looked so much alike that in Korea, some US pilots, eager to bag a MiG, accidentally shot down a number of Sabres.

'The MiG-15 surprised the hell out of us', said US pilot Robert van der Linden. Compared to the F-86 Sabre, 'the MiG was faster, could outclimb it, and had more firepower'. And the Sabre pilots knew it.

'MiG Alley' was area where many of the jet-fighter dogfights took place. It lay over the north-west of North Korea, where the Yalu River, the border between China and Korea, empties into the Yellow Sea.

All the old piston-engined fighters were left standing, with one rare exception. On 8 August 1952, Lieutenant Peter 'Hoagy' Carmichael of 802 Squadron was flying his Hawker Sea Fury near Chinnampo,

---

8 Swept wings were also fitted to the rocket powered Messerschmitt Me 163 Komet. This astonishing aircraft had set an unofficial world speed record in 1944 of 702mph (1,130km/h). It was so fast and could climb to altitude so quickly pilots had to be put on a special low-fibre diet to avoid gas expanding in their gastrointestinal tract. The combined effect of loose bowels; due to the speed, and flatulence; due to the altitude, could lead to a loss of control.

North Korea: 'Eight MIGs came at us out of the sun. I did not see them at first, and my No. 4, "Smoo" Ellis, gave a break when he noticed tracer streaming past his fuselage… One MiG came at me head on. I saw his heavy tracer shells. I fired a burst, then he flashed past me.'[9]

Carmichael was credited with shooting down this MiG-15, making the Sabre-engined Hawker Sea Fury one of the few propeller-driven aircraft to shoot down a jet during the Korean War.

The MiG-15 jet fighter went on to shoot down at least 142 UN Command allied aircraft during the Korean War – the vast majority of them being US aircraft. Analysis of ground control traffic in June 1952 concluded that more than 90 per cent of MiGs engaged in air operations over North Korea were being flown by Russians masquerading as North Koreans, reading out Korean-language radio acknowledgements from a cheat sheet taped in the cockpit.[10] These turned into Russian shouts during the stress of combat. MiG-15 pilot Porfiriy Ovsyannikov was on the receiving end of the B-29's guns: 'When they fired at us, they smoked, and you think, is the bomber burning, or is it machine gun smoke?'

This direct Soviet participation was deliberately ignored by UN Command, lest this became a *casus belli* to escalate the war to include the Soviet Union, and potentially cause a second atomic war, just five years after the first one.

---

9 *Take Off* magazine, Part 84

10 Budiansky, Stephen, *Code Warriors*, New York, Alfred A. Knopf (2016), pp. 140–141

Chapter 5

The World's First Jet Airliner

The de Havilland Comet was an extraordinary aircraft. On its maiden flight in 1949, it was the most spectacular commercial aircraft ever built: here there were no visible engines, no thrashing propellers, no roaring vibrations. It was just stunningly elegant. It was also the world's first commercial jet airliner, the first jet airliner to cross the Atlantic and the Pacific, and it inaugurated the first regular jet-powered transatlantic service between London and New York.

Seventy years on, jet airliners still look much the same as the Comet, and when you pass a British road sign warning of aircraft, the silhouette is, of course, that of a Comet.

On 1 August 1942, Winston Churchill travelled to Moscow via Cairo for a conference with Stalin. He had an uncomfortable ride in the freezing bomb bay of a Liberator bomber, wearing an oxygen mask and alternating between the frost of altitude and the heat of the desert. This trip brought home to him the lack of a modern British airliner.

The aviation press thought the same: 'The whole British Empire at the present time has an operational fleet of transport aircraft, comprising conversions, makeshifts and cast-offs, totally inadequate to represent the Empire in serving the air routes of the world in the peace to come. Have we to rely upon other nations to do it for us?

The British aircraft industry is equal to the task. The Government should decide this vital question at once.'[1]

In Parliament, Walter Robert Dempster Perkins MP, first parliamentary secretary to the Ministry of Civil Aviation, rose on to his hind legs: 'I foresee that, as soon as the armistice comes, there will be a race between the Americans and ourselves to control the airlines of the world.'

Perkins, a Conservative MP, had clearly been primed by Churchill. The British government, he suggested, must agree 'which spheres of influence might belong to America and which to Great Britain. Otherwise, there will be friction. We might even have another Boston Tea Party.'[2]

Spheres of influence had been a preoccupation of the British since Persia (now Iran) had been divided up into two such spheres to provide a buffer zone between the British and Russian empires. There were others in Siam (now Thailand) and China. The concept might seem strange to Britons now, when the only such sphere is over a few scattered islands, but it suggests that Churchill still clung to the idea that the British Empire could somehow be revived, whereas the truth was that, between them, the Germans and Americans had made quite sure that no such thing was going to happen, thank you very much.

One wonders how much the British public really cared about the colonies by that time. In 1883, the historian J. R. Seeley wrote that the empire seemed to have been acquired 'in a fit of absence of mind'. By that, he didn't mean to suggest a sense of effortless superiority; Seeley was being sarcastic, and his point was that English history textbooks of his day didn't discuss the empire at all. One recent

---

1 *Flight* magazine (24 December 1942)

2 Perkins married the widow of Air Commodore Sir Nigel Norman, yet another victim of an air crash. In this occasion it was a Lockheed Hudson, which lost one of its two Wright Cyclone engines on take-off from RAF St Eval. The old pilot's adage about twin-engined aircraft still obtained: the good engine just took him to the scene of the accident.

historian[3] pointed out that the British public knew little about the colonies and cared even less. The percentage of boys at public schools going on to colonial administrative careers was only between 2 and 5 per cent, and parliamentary debates about the empire were thinly attended. In 1860, just 10 per cent of the House of Commons' time was taken up by imperial affairs.

Only the upper echelons of government and certain wealthy businessmen seemed to be concerned with the colonies.

And so, in March 1943, in the middle of a world war, with a surprisingly confident expectation of victory, the Brabazon committee assembled. Their task was to determine which types of British civil aircraft were to be built after the war.

It might seem curious to us now in our post-colonial world, but the Brabazon committee's main concern seemed to be with providing fast mail links to the far-flung colonies of the British Empire and Commonwealth. Only small numbers of passengers were envisaged – military top brass, the wealthy or colonial administrators. Command of the seas had served the British Empire well for two centuries; couldn't command of the skies provide the same advantage? Steam engines had served well in the past, having been the 'Engines of Empire'; could the new jet engines serve even better for the future?

The committee's chairman, Lord Tara of Brabazon, was one of Britain's foremost aviation pioneers. He was a Conservative MP and Minister of Transport and Minister of Aircraft Production during the Second World War. He had been forced to resign in 1942 for expressing the widely felt hope that Germany and the Soviet Union, then engaged in the Battle of Stalingrad, would destroy each other. Since the Soviet Union was still fighting the war on the same side as Britain, this sentiment had to, officially, be frowned upon.

---

3 Porter, Bernard, *The Absent-Minded Imperialists*, Oxford University Press, Oxford (2004). Porter concludes that Britain was 'a less imperial society than is often assumed' and he draws a parallel with today's United States, where half of society is evidently not imperialist at all. Porter clearly had no enthusiasm for empires.

On 2 May 1909, Brabazon had been the first resident Englishman to make an aeroplane flight in England and he was the first to qualify as a pilot in the United Kingdom, with the first licence. His car's registration plate read 'FLY 1'. He befriended the Honourable Charles Rolls, another pioneering aviator, and once carried a live piglet in a washing basket strapped to the wing of his aircraft. This was to prove that pigs could fly.

Brabazon wanted a large propeller-driven airliner for transatlantic routes. Vested interests were also present on the committee in the form of Vickers, which wanted to build a turboprop-engined plane for European routes, and Geoffrey de Havilland, who wanted a jet-powered high-speed transport. At that time, his company was, of course, involved in developing Britain's first jet-engined fighters. Unfortunately, not all the five types of aircraft chosen fitted well with the post-war world.

Type I became the Bristol Brabazon, a huge double-decked aircraft named perhaps sycophantically after the committee and its chairman. Although it was one of the biggest aircraft ever built and at first it had no less than eight Bristol piston engines, it only seated 100 passengers, each of whom enjoyed the same space as the interior of a car. The plane boasted a bar, a smoking lounge, a dining area and even a 32-seat cinema. The village of Charlton near Bristol was compulsorily purchased and demolished to lengthen the runway needed to accommodate this monster, with the residents being relocated to Patchway.

But the Brabazon was a flop. No one bought it: 'The spec wasn't correct for post-war flying. The people who wrote the specs… conceived of an aeroplane with all this comfort, bunks, and a great dining room to eat in. And, of course, come the day, that wasn't what the airlines wanted. They wanted to ram as many passengers as possible into the tube and give them lunch on their laps.'[4]

---

4 Kaplan, Philip, *Big Wings: The Largest Aeroplanes Ever Built*, Pen and Sword, Barnsley (2005)

In a way, the Brabazon was 20 years too early, foreshadowing the Boeing 747 Jumbo jet in size. The efforts that went into its construction were not wholly wasted, as Bristol went on to produce the Bristol Britannia, which used the same facilities built at Filton. The Britannia was powered by four of the same Proteus turboprops developed for the Brabazon, and it weighed almost 40 per cent less than its predecessor. It could carry the same number of passengers across the Atlantic at a cruising speed more than 100mph (160km/h) faster. In the end, Bristol delivered 85 Britannias, plus a further 72 built under licence in Canada.

The Brabazon Type IV proposal was more of a success than the Type I Bristol Brabazon, as it became the de Havilland Comet. This was for a pressurised jet-powered transatlantic 40-seat mail plane. It was an ambitious suggestion.

It must be remembered just how awful air transport was before the jet airliner. Before you boarded, you knew that the highly flammable petrol that you could smell made the frequent accidents even more dangerous. Then there was the racket: 'When the day was over my bones ached, and my whole nervous system was wearied from the noise, the constant droning of the propellers, and exhaust,' complained one passenger. One pilot, Ernest K. Gann, said the worst thing was the smell: 'The airplanes smell of hot oil and simmering aluminium, disinfectant, faeces, leather and puke.'

If the turbulence, thin air, lurching and stench made you need to vomit, there were no sick bags strong enough until plastic-lined bags appeared in 1949. One solution was to open a window and try to vomit outside because the aeroplanes flew low enough not to require pressurisation or sealed windows. 'If you stuck your head out to do this,' said Gann, 'you might find someone a few seats ahead with the same idea, and you could receive his blast in your own face.' In many aeroplanes, there were holes in the floor so that after a particularly vomit-inducing flight, the seats could be hosed down.

Jet airliners would make air travel more pleasant in at least three ways: firstly, jet turbines spin smoothly instead of reciprocating and vibrating like piston engines and propellers. Secondly, jet planes fly at twice the altitude of piston planes, far above all the turbulence and bumps of bad weather. Thirdly, they fly twice as fast, cutting flight times in half. Compared to the old sick-making piston aircraft, jet travel was going to be a delight.

Once again, a child's toy was to inspire a lifetime's devotion to flight, as did the toy aeroplanes of Frank Whittle and the Wright brothers.

The young Geoffrey de Havilland flung a toy parachute out of the attic window of his father's parsonage and watched it fall. He had found it stuffed inside a Christmas cracker, and he repeatedly dropped the parachute out of the window and raced down the echoing stairs to retrieve it from the garden. One day though, a miracle happened. 'Instead of falling directly to Earth,' he remembered many years later, the parachute 'started to rise, and went on rising until it passed over the stables in the yard. This was, of course, due to a local up-current of air, but this seemed to have some special significance.'[5]

Geoffrey became fascinated by flight and motorcars, much to the despair of his reverend father,[6] author of the snappily titled *An Outline of the History of Ancient and Modern Rome; Comprising an Account of Italy, from Its Most Remote Antiquity to the Present Time, and Embodying the History of Christianity, from Its Earliest Date.*

Young Geoffrey worshipped speed instead. As soon as they became available, he and a friend hired one of the new motorcars with a driver and went for a spin. 'We had scarcely, in the popular phrase,

5 de Havilland, Sir Geoffrey, *Sky Fever: The autobiography of Sir Geoffrey de Havilland*, Hamish Hamilton London (1961), p. 96

6 Actresses Olivia de Havilland (*Gone with the Wind*) and Joan Fontaine (*Rebecca*) were Geoffrey's cousins.

annihilated space, but after that short drive I knew that my future life lay in the world of mechanical travel.' The notion of shortening distances with higher speeds, he said, 'gained a hold which was never to relax through all my working life.'

Geoffrey de Havilland went on to become an engineer, then an aircraft and aero-engine designer. His 5-litre water-cooled flat-four 'Iris' engine powered a 1909 de Havilland biplane of his own design. During the First World War, he modified his BE-2c biplane design specifically to counter the menace of Zeppelin airships.

In Germany, children were being taught to sing 'Fly, Zeppelin! Help us win the war, England shall be destroyed with fire. Zeppelin, fly!' And indeed, the giant balloons were a menace, killing 2,000 civilians and causing enormous damage. The Germans were triumphant, and one newspaper proclaimed:

> It has come to pass – that which the English have long feared and have repeatedly contemplated with terror. The most modern air weapon, a triumph of German inventiveness and the sole possession of the German military, has shown itself capable of crossing the sea and carrying the war right to the sod of old England!

After 18 months of unopposed bombing, a Zeppelin was at last shot down by a de Havilland BE-2c. Five more Zeppelins followed, and de Havilland was proclaimed a saviour. No less than 3,500 versions of his BE-2 were built all over the empire, and in the United States 95 per cent of wartime aircraft were built to a de Havilland design.

Then, as we have seen, in 1915 the tables turned when the German Fokker Eindekker aircraft appeared over the trenches and the BE-2 was hopelessly out-classed. Now there was public indignation and parliament ordered a judicial enquiry into the dreadful performance of British aircraft. De Havilland came under huge public pressure and, although exonerated, he suffered a nervous breakdown just before the war ended.

Slowly he recovered and regained his energy. But his depression had been a warning: 'I was able to do some of my best and most strenuous work in design and flying in the years that followed. But the illness has been a severe warning… that there were definite limits beyond which I must never again trespass.'[7]

His aircraft explored their limits successfully, though, and in 1920 he formed a successful aircraft company, which at first built the popular 'Moth' series of biplanes. A particularly sweet triumph was the winning of the World's Greatest Air Race, the 1934 England–Australia MacRobertson Air Trophy. Completing its maiden flight just six weeks before the race, the twin-engined DH-88 'Comet' was a stressed-skin cantilever monoplane with an enclosed cockpit, retractable undercarriage, landing flaps and variable-pitch propellers. It was made of wood for lightness.

It had a top speed of 237mph (381km/h, 206 knots), and it beat a Boeing 247, which came in third, with an elapsed flying time of 92 hours, 55 minutes. A DC-2 came second in 90 hours, 13 minutes. But the de Havilland Comet flew the course in only 71 hours. For Geoffrey de Havilland, it was an international triumph. So proud was he that he revived the Comet name for his jet airliner 15 years later.

During the Second World War, the wooden de Havilland Mosquito became renowned as the most versatile warplane ever built.

Next, Geoffrey wanted to get into the new jet engine technology. As we have already seen, de Havilland's friend Frank Halford had built a centrifugal jet engine based on Whittle's design but had improved it somewhat with a 'straight-through' configuration. This first flew on 5 March 1943 in the Gloster Meteor, and then on 26 September in the de Havilland Vampire, Britain's second jet fighter.

It was then that de Havilland purchased Halford's company, asking him to act as chairman of the new de Havilland Engine Company,

---

7 de Havilland, *Sky Fever*, p. 96

with the engine name changing from H-1 to 'Goblin', while the new H-2 design became the 'Ghost' (as Rolls-Royce used river names for its jet engines, so de Havilland used supernatural apparitions for its jet and rocket engines).

Tragically, de Havilland's youngest son, John, was killed during testing one of his father's aeroplanes. He died in an air collision involving two Mosquitos in 1943.

The Brabazon committee accepted de Havilland's proposal for the Type IV jet-engined 40-seat air liner and, in February 1945, awarded a contract for the development and production of what was to be called the de Havilland Type 106 Comet. The war then was by no means over: German forces were still holding out west of the Rhine.[8]

The British Overseas Airways Corporation (BOAC), whose representatives were on the Brabazon committee, liked the look of the Type IV's specifications and initially proposed buying 25 aircraft. However, in December 1945, when a firm contract was drafted, BOAC's order was changed to ten Comets.

The Type 106 was going to be the riskiest of all the five Brabazon types in terms of its advanced design, and a huge financial outlay was involved. It was so technically advanced that de Havilland had to undertake the design and development of both the airframe and the engines.

To begin with, the fuselage was to be pressurised. Cabin pressurisation is necessary as an aircraft climbs above heights of around 10,000ft, otherwise passengers start to suffer from hypoxia (lack of oxygen), altitude sickness and barotrauma (pain in the ears due to trapped gases). As they descend again, the unfortunate

---

8 At this late point in the war, tiny Paraguay bravely took the opportunity to declare war on Germany and Japan, an odd decision, as so many fleeing Nazis would end up in the country.

passengers may then suffer from decompression sickness. All of these symptoms can be avoided by pumping the aircraft fuselage up with compressed air to simulate an altitude of around 8,000ft. De Havilland designers had to work out how to pressurise their Type 106's fuselage safely. Previous piston-engined plane designers had experimented with oxygen-enriched air in the cockpit up to 40,000ft, but the pilots had suffered from enlarged hearts, and, of course, there was an increased fire risk.

The first passenger-carrying aircraft to use the pressurised system such as the Boeing 307 and Avro Tudor were described as having 'supercharged cabins', a term that could surely catch on today. They used mechanically driven pumps, but de Havilland invented a way to pressurise the cabin using air bled off the compressor stage of the jet engines.

The problem was that the cruise altitude could be as high as 40,000ft with a cabin pressure equivalent to 8,000ft, requiring pressurisation twice that attempted on any previous airliner. The difference between the inside and the outside of the cylindrical fuselage was going to exceed anything attempted before. This was going to be like flying in a tense blown-up balloon with a very thin skin.

Other technical advances that had to be met were 20° swept-back wings mounted low enough to pass below the pressurised cabin, four new jet engines buried in the wing roots, cabin heating good enough to cope with an outside temperature of -70°C, fuel tanks integral with the wings, power-assisted flying controls and a tricycle undercarriage. Any one of these might have stumped an ordinary designer, but Ronald Bishop was no ordinary designer: he had been responsible for the Mosquito.

His design team fretted particularly over the strength of the pressurised cabin. Like a submarine's pressurised hull, this must not be allowed to fail. At 40,000ft, there was going to be a pressure differential of 8¼ pounds per square inch (psi), so they designed in a strength factor of 2.5 (20½ psi). They tested one fuselage to destruction using air pressure and it exploded like a bomb, destroying all the evidence. So, they built a water tank and burst the fuselage more gently, using water pressure, thus localising the damage. 'We treated the windows as a special case,' said Bishop later, 'and tested them to a safety factor of ten (eighty-two pounds per square inch).'

The Comet windows were not made round, like a ship's portholes. Perhaps to distance the world's first jet airliner from old-fashioned nautical transport, they were made a different shape. There is a popular misconception that the Comet's windows were square, with sharp corners, but in fact they were rectangular, with heavily rounded edges. On top of the cabin, instead of drag-inducing aerials, the Comet had two square automatic direction finding (ADF) 'windows', which were not really windows as they were not transparent. These were made of electrically transparent glass fibre reinforced plastic squares – high-tech for the time.

Chief designer Ronald Bishop embedded the jet engines in the roots of the wings because this configuration avoided the drag of podded engines and therefore allowed for a smaller fin and rudder, since the hazards of asymmetric thrust were reduced. In this, he probably had in mind the difficulty of landing de Havilland Mosquitos with one of the Merlin engines disabled: the thrust from one side was exacerbated by the drag from the other.

Jet engines are extravagantly thirsty at low altitudes, and the entire economic success of the Comet was to depend on it flying high where the air was thin and offered less resistance. Therefore, the fuselage had to be built as strongly yet as lightly as possible. Even then, only 10 per cent of all-up weight was available as payload. New building techniques were developed: the aluminium skin was only as thin as a postcard, and it was attached to the skeleton frame with adhesive instead of rivets. Rivets add weight and air-skin friction and require holes that can leak. The holes for any rivets that *were* needed were punched for speed, not drilled. The balsa-wood Mosquitos had been stuck together with Casein, a milk-based glue, but this time de Havilland used a special metal-to-metal cement known as Redux. Redux was a resin-based adhesive and this was most advanced for the 1940s.

Because of the thirst of the Ghost engines, 40 per cent of the Comet's weight would be fuel, and so the wings themselves would have to become integral fuel tanks, another Comet first that has become a standard feature in jet aeroplanes today.

The powered controls were another innovation, using hydraulic pressure to move the control surfaces rather than the pilot's muscles. There were difficulties with the correct amount of feedback: as any car enthusiast knows, unpowered steering and brakes provide more information back to the human operator. Powered controls can feel dead and can move too easily and quickly.

Although built for twice the speeds of rival piston-engined aircraft, the wing-loadings on the Comet were actually lower than other piston-engined airliners and it had a reasonably slow stalling speed. This meant it could descend slowly through low cloud, perform a standard circuit and make a landing in the manner of a normal aircraft. This was an important selling point.

Perhaps the most surprising feature of the Comet was the plan to use rocket motors to assist take-off from the 'hot and high' outposts of the empire that so preoccupied the Brabazon committee. It was known that due to air density and temperature, tropical high-altitude airfields such as Nairobi's Eastleigh Airport at 5,370ft reduced jet-engine thrust by as much as 25 per cent. And the Ghost engines were only just powerful enough.

The de Havilland DH Sprite rockets were fuelled with hydrogen peroxide and provided the enormous thrust of 5,000lbf: the same as a jet engine. A pair of Sprite rockets would therefore be the equivalent of another pair of Ghost engines added for take-off. They would only fire for around 12 seconds, but that was enough to assist the aircraft into the air. The Sprite rockets would also have been used for the forthcoming V-Force nuclear bombers in order to get their heavy loads into the air, and, in fact de Havilland regarded the 166 Sprite rockets that were manufactured as standard production items, supported by its service department alongside piston and turbojet engines. Imagine ordering a couple for your car…

The idea of attaching rockets to a civil airliner might seem unwise to us now: what if a pilot accidentally sat on the button while facing the terminal? In the end, the increases in power from the next generation of axial jet engines made rockets unnecessary.

The Comet might have been excused for looking like a lash-up with all this attention paid to high technology, but, in fact, the resulting aircraft was a sleek, clean, elegant design, with a smooth nose and no sign of ugly engines. The BOAC livery was the best of British reserve: white and silver with a thin blue line along the length of the windows. The blue extended around the windscreen, rather like the bandit's mask of a swallow. The Comet really was one of the most beautiful aeroplanes ever built.

The swallow of summer, she toils all the summer,
A blue-dark knot of glittering voltage,
A whiplash swimmer, a fish of the air.[9]

The de Havilland Ghost turbojet engine was specially developed for the Comet. It was still a centrifugal-compressor type engine and Frank Halford had scaled up his Halford H1 using ten larger combustion chambers in place of the H1's 16 small ones, using bifurcated split intakes that were fed by each diffuser duct. Jet engine designers were still feeling their way with the new technology, but clearly there were future gains to be made. The new Ghost was more powerful than its predecessor, with a maximum thrust of 5,300lbf (24kN) at

---

9 Hughes, Ted, 'Work and Play' (1961)

10,350rpm for take-off. This would equate to a propeller engine of 12,000hp due to the propeller efficiency of around 53 per cent.

However, it still wasn't really powerful enough for a heavily loaded airliner. Four Ghosts had to be used. The low pressure ratio of 4.6:1 also meant that it burned a lot of fuel. However, the engine was soon to set an important new world record.

De Havilland's chief test pilot was John 'Cat's Eyes' Cunningham, who had been an RAF night-fighter ace during the war. He had been given his nickname by the press to disguise the successful use of radar, and he was a notably cool customer. On one occasion, just before the war, he had been chased through the sky by his own unpiloted aircraft, along with Geoffrey de Havilland. The two men were testing a Moth Minor aircraft for safe response to spins. The aircraft was put into a right-hand spin, the nose shot up, the engine stalled and the Moth Minor entered an irretrievable flat spin. The two men looked at each other, jumped and their parachutes opened. They then looked around for their aeroplane. Unfortunately, the lightened Moth Minor had dropped its nose, recovered from the spin and the wind-milling propeller had restarted the engine. The aircraft started to circle, pursuing the two parachutists in ever-decreasing circles around the sky. Luckily, it descended slightly faster than they did, and it crashed onto the ground just below them. Cunningham calmly took photographs of the parachutes and the wreckage, unflustered by the experience.

The de Havilland Ghost engines were chosen for the Comet because the more powerful Rolls-Royce Avon engines with 7,000lbf (31kN) weren't yet ready. This choice probably saved two years of development time. It was established that the Goblins were not affected by the extremely cold conditions of high altitude, nor had they been damaged after swallowing large quantities of ice during testing, including chunks weighing several pounds each. The robust impeller of the centrifugal compressor had successfully shattered and digested the chunks, whereas the delicate compressor blades of an axial type of turbojet might have been damaged. This was one of the reasons for the choice of the Ghost, even though it was realised that the axial jets that were coming were lighter, more powerful, more fuel-efficient and more compact.

The Ghost was first run up on the test bench just after the end of the war in Europe. Geoffrey de Havilland was wasting no time. For flight testing, two Ghosts took to the air on 24 July 1947, on an Avro 691 Lancastrian, VM703. This was a piston-engined airliner converted from a Lancaster bomber and, as such, usually mounted four Merlin engines.

An Avro Lancastrian provided one of the most enduring mysteries of aviation history, and the story underlines the discomforts and dangers of piston-engined flight. Just a week after the Ghost's first flight, a Lancastrian on a flight from Buenos Aires, Argentina, to Santiago, Chile, disappeared without trace. Air searches failed to find a crash site, and the only clue was a mysterious Morse-code message: the word STENDEC, repeated twice. The passengers were a mixed bunch, straight from the pages of an Agatha Christie novel: a Palestinian jewel dealer, a British diplomatic courier, a young woman and a mystery German. Theories abounded: ex-Nazis on the run, a jewel robbery, top-secret documents and a UFO abduction.[10] Fifty years went by with no trace, until two Argentinian mountaineers attempting the 21,560ft (6,570m) Mount Tupungato in the Argentine Andes found a Merlin engine, a still-inflated aircraft tyre and a severed female hand. The nails were carefully manicured. More pieces of the Lancastrian were emerging from the glacier.

---

10 The word STENDEC was corrupted into Stendek and became the name of a Spanish UFO magazine.

One theory held that the crew had become confused as to their location due to a high-altitude jet stream acting as a headwind. Believing they had cleared the Andes, they descended too soon, hitting a vertical ice wall which then avalanched, hiding the wreckage in the glacier. This was then duly disgorged by the slow-flowing ice after 50 years.

However, in 2000, an Argentinian Air Force inquiry concluded the plane had crashed because of severe weather and cleared the pilot, Captain Reginald Cook, of any blame for the deaths of the five crew and six passengers. Major Luis Estrella said a heavy snowstorm forced the decorated RAF veteran to veer off course. 'They were unable to correct their positioning because they were flying in very cloudy weather,' he said.

Gremlins or aliens were probably not responsible.

Meanwhile, back at Hatfield, the test Lancastrian was now testing not only the two Ghost jet engines but also a pair of rockets. There is a memorable photograph of the venerable Lancastrian VM703 taking off like a scalded cat, trailing clouds of smoke from Merlins, Ghosts and rockets.

The Lancastrian's service ceiling, however, was limited to 25,000ft, and further tests were required above 40,000ft. By then, de Havilland had its new single-engined jet fighter, the Vampire, under development. Fitted with the Ghost engine and with specially extended wings, on 23 March 1948 Cunningham got the Vampire up to 59,446ft, a new world altitude record. The Ghost engine performed well, running in an air pressure of only 0.02 of an atmosphere. This success would help Vampire sales and gave further encouragement to the team working on the Comet airliner. Workers at de Havilland could justifiably feel that they were the world's finest builders of aircraft.

Now they had to get on with building the world's first airliner. Great pains were taken with the new techniques of bonding the stringers or longitudinal braces to the thin aluminium skin. De Havilland had always been an innovative builder: during the recent war, it had pioneered the use of radio-frequency heating to accelerate curing of the adhesive on the Mosquito: a forerunner of the microwave oven. The Comet wings were fitted so exactly to the fuselage that when measurements were taken from the tips of the 110ft-span wings to the rear fuselage cone, there was only a difference of 0.4 of an inch (10mm) between them.

It was time for the first test flight. Chief test pilot Cunningham flew without the number two test pilot in case he was killed. He reported:

I taxied back to the grass at the east end of the runway and prepared for take-off. The aircraft came off the ground in about 500 yd, and on reaching about 50 ft high I selected u/c [undercarriage] up. As the speed was increasing rather quickly, I throttled down to about 9,500-9,700 r.p.m. and 135kt. About this moment the windscreen-wiper blade on the starboard side started to rotate: a position was found where it could be parked out of sight for the remainder of the flight.

The ailerons appeared rather light, but very effective. The flap was taken off in two stages, and very little trim change was noted. A climb was made towards a clear patch of sky at 9,750 r.p.m. and 240 knots. Jet pipe temperatures of all engines showed 580°C and rear bearing 100°C. Engine oil temperatures were 63°C. During the climb at 200 knots, I found all controls very powerful and highly geared; the spring centring of all controls produced a pronounced jerk throughout the machine on releasing the control after slight displacement.

On reaching 10,000ft I lowered 20° of flap and reduced speed to 100 knots and lowered full flap. There was very little change of trim noticeable… Wheels were next selected down, when it was noticed that no lights were showing, although the mechanical indicators showed 'Locked Down'. While Waters (the electrician) was checking the circuit breakers I reduced speed to 80 knots, where there was still plenty of control and no buffet. On replacing the circuit breakers, I noticed there was no green light for the port main wheel, although the mechanical indicator showed 'Down'. Wheels and flaps were selected up and a descent made to 3,500ft with the inboard engines throttled back. At 2,500ft, cruising at 8,700 r.p.m. on all four engines, I was indicating 240 knots. A gradual descent was made back to the airfield, during which it was extremely hard to lose speed. After a low flypast at 150-160 knots down the runway, a wide circuit was made, and the final approach to land was completed at 100 knots. The landing and touchdown were perfectly straightforward, and very little braking was required owing to a fairly strong wind.

This would count as a successful first flight, although nerve-wracking for the average pilot. A lifelong bachelor, Cunningham always evaded reporters' questions about his romantic interests by pointing to the Comet and saying: 'That's my wife. She's beautiful and drinks paraffin.'

All was well, and the Ghost jet engines proved to be largely trouble-free throughout their service life in the Comet. A pilot used to piston engines would have smiled at the thought of *reducing* revs to 9,500–9,700rpm, when few aero engines could reach even 3,000 revs. Also remarkable was the simplicity of the jet engine controls in the cockpit compared to those of a piston engine. Furthermore, as the passenger brochure pointed out, there was no lengthy warming-up of engines necessary: 'A polite swish... and no vibration. A flight in a Comet Jetliner is a memorable, oddly soothing sensation. It takes but a minute to start the four jet engines, buried in the depth of the wings, before this low-wing monoplane taxis to the runway. No protracted warm-up is necessary.'

De Havilland pressed on with production, and on 2 May 1952, the world's first jet airliner on the world's first jet airliner service took off from London Airport, known today as Heathrow. The Comet, registration number G-ALYP, or 'Yoke Peter' in the phonetic jargon of the day, was bound for Johannesburg, via Rome, Beirut, Khartoum, Entebbe and Livingstone. The BOAC publicity brochure writer caught the moment:

> Easily, without apparent effort or the fury of flaying propellers, the Comet slides smartly into the air. Climbing 35–40,000 feet, she levels off on a rock-steady flight above the highest clouds in a dazzling, almost indigo sky. The Comet is remarkably quiet and free from vibration. Little outside noise is perceptible. Lack of vibration, and the absence of any sign of an engine, propeller or other moving part, completes the illusion of being fixed rather than moving in space. There is no sensation of distance dropping behind at eight miles a minute. You are, in effect, sitting in a pleasant room above a painted scene.

Strong headwinds over the Alps made them nine minutes late in Rome, where they refuelled. Those early jet engines were immensely thirsty, and so the stages had to be short and the fares commensurately high. They made up time on the next leg and were 15 minutes ahead of schedule at Beirut. The BOAC writer conjured up the heat, the night and the moonlit romance of early jet travel:

> While the clattering kerosene bowsers once more disgorged their thousands of gallons into the great wing tanks, Captain Majendie handed over to the slip crew who were to take the service on through the night. And so to Khartoum, silvery in the waning moonlight.

More fuel, another fresh crew, a stroll for the yawning passengers, and they were off again, the brazen sun tipping the rim of the desert as the Comet winged southwards to Entebbe on the edge of Lake Victoria; then on to a last refuelling stop in the midday furnace heat of Livingstone. Climbing away over the Victoria Falls, there were still a few minutes in hand, and they dawdled along the last leg of the journey, making wide turns to kill time. When Yoke Peter finally touched down at Johannesburg, twenty-three and a half hours after leaving London, it was just two minutes ahead of schedule.

The flight was a delight. The tired but happy passengers reported a smooth journey, flying above bad weather en route. The Comet flew twice as high and twice as fast as the old piston-engined planes, with half the noise; this was the new way to travel, and these travellers were the new Jet Set. For a brief, euphoric moment, Britain held the world lead in jet airliners.

The news was good from the technical side, too. The cost of maintaining a conventional piston-engined airliner in the 1950s was calculated to be around 40 per cent of the total operating cost. Added to this was the regular extra cost of engine trouble, which upset flying schedules and gave rise to unexpected expense, often involving the flying of replacement engines to remote airfields. De Havilland, keen to offset the impression of high fuel bills, made much of the ease and cheapness of servicing the new jet engines:

Consider, as an example, the inspection of the ignition system. For the piston engine: change up to as many as 56 sparking plugs[11] (which

---

11 They were clearly referring to the Pratt & Whitney R-4360 Wasp Major, an American 28-cylinder four-row radial piston aircraft engine used in the Boeing 377 Stratocruiser. It was known as the 'Corncob'.

may be inaccessible and may involve removing baffle plates), check magneto, contact breaker gaps and the insulation and continuity of the plug leads and switches.

For the Comet: inspect the igniter plugs, which were used for starting only, and make a function check. Such a minor inspection for one engine (including, of course, a daily inspection as described above) requires about two man-hours, so that approximately eight man-hours suffice to complete a minor inspection of all four.

Another encouraging fact emerged. Having no experience other than military jet operations, at first de Havilland had to calculate the service life of the Ghost under military operating conditions, which are extreme. Civil airline pilots, however, were gentler with the engines, opening the throttles more slowly, which avoided rapid changes in temperature of the combustion chambers, the turbine blades and the jet pipe. The service life, initially expected to be 500 hours, doubled.

Only one significant problem emerged with the Ghost engine. After many hours of running, small cracks began to appear in the impeller turbine of the centrifugal compressor. These were being caused by a high-frequency vibration, and the cure was one the Germans had employed: slightly cropping the blades of the impeller to shift the frequencies of vibration outside the dangerous range.

If the Comet's jet engines did have to be changed, it was a quick operation compared with a multi-cylinder piston engine. The Ghost engines may appear to have been buried in the wind roots but, in fact, they were accessible from underneath at a convenient head-height. The fuel-system components and all other accessories were mounted forward of a fireproof bulkhead, and each engine was mounted on a pair of trunnion bearings with one steadying tie bar. De Havilland's wartime experience showed in the quick-release pipes, cables and controls: the throttle controls could be released and reattached without having to readjust. The jet pipe could be released and slid back on runners while the engine was lowered using a chain hoist attached from above through a hatch in the wing. All four Ghosts could be changed in 90 minutes, employing three engineers for each pair of engines.

Prince Philip returned from the Helsinki Olympic Games on board a Comet on 4 August 1952, and the Queen Elizabeth, the Queen Mother and Princess Margaret were guests on a special Comet flight on 30 June 1953 hosted by Sir Geoffrey and Lady de Havilland. During the flight, the Queen Mother took the controls.

If the British royal family flew on the new jet, it had to be safe. Surely?

Airline executives were impressed. On 20 October 1952, de Havilland announced that Pan American World Airways (Pan Am) had bought three Comet Series 3 jet airliners. Juan Trippe, the founder and chairman of Pan Am, was to become the jet airliner's greatest proponent. For the first time in aviation history, a British aircraft had been chosen by an American airline operator and, for a brief moment, Britain held the world lead in jet airliners.

Chapter 6

Downfall

T‍hen two incidents happened to the Comet, both attributed to pilot error. On 26 October 1952, a BOAC Comet was about to take off from Rome's Ciampino Airport. Runway visibility was 5 miles (8km), but the pilot could not see the horizon. The windscreen wipers were on; the elevator, aileron and rudder were all set at the neutral position and the flaps were lowered to 15°. The engines were set to full power; their revs were 10,250rpm and the temperatures all correct. The brakes were released, and the Comet started to roll. At 112 knots, the pilot, Captain Harry Foote, pulled back on the control column and the aircraft took off. This lifting of an aircraft's nose is termed 'rotating'. Immediately though, the port wing dropped, and severe buffeting occurred – the sign of a stall. Captain Foote had the impression that the engines had lost power. Air speed was not increasing. The aircraft landed heavily back on its undercarriage and ploughed off the runway, wrecking itself in rough ground. No one was hurt.

The investigation found a series of tail scrape marks down the runway, and the inquiry blamed Captain Foote for 'an error in judgement by the captain in not appreciating the excessive nose-up attitude of the aircraft during the take off'. There was a quote from the Comet operating manual: 'At 80 knots the nose should be raised until the rumble of the nose wheel ceases. Care should be taken not to overdo this and adopt an exaggerated tail-down attitude with consequent poor acceleration.' De Havilland said that if the Comet was rotated or tilted to 9°, there would be a partial stalling of the wing, reducing acceleration and bringing on a buffet that Captain

Foote felt and tried to correct. With an 11.5° rotation, the tail would hit the ground.

Captain Foote, although highly experienced, was accused of 'over-rotating' and demoted to flying freighters. He was mortified.

Then, on 3 March 1953, at Karachi, Pakistan, during a night-time take-off, a similar accident occurred. The heavily loaded Canadian Pacific Airlines Comet was seen to have an exaggeratedly nose-up attitude during the take-off run, the tail hit the ground several times and the aircraft refused to become airborne. It used up all the runway and the overrun, hit an earth bank at 138mph (222km/h) and exploded into flames. All 11 on board were killed. This Comet was the first passenger jetliner involved in a fatal accident. At the inquiry, again the pilot was blamed.

On 2 May 1953, exactly one year to the day after the first triumphant Comet flight, members of the British expedition on Mount Everest were making a preliminary reconnaissance of the Lhotse Face preparatory to the first ascent of the mountain later that month.

A few hundred miles to the south, just as the British mountaineers were returning to their tents at 1630hrs, a de Havilland Comet took off from Calcutta's Dum Dum Airport. On board were passengers invited to the new queen's coronation. The weather was monsoony, and an airfield weather report had been issued just before the Comet had landed, warning of a thunderstorm approaching with squalls reaching 50 knots. The thunderstorm was reported to have 'very strong vertical up draughts'. Speaking to Calcutta Area Control at 1632hrs, the Comet reported that its estimated time of arrival at Delhi was 1850hrs and it was climbing to 32,000ft. At 1653hrs, the Comet was heard calling

Delhi, but when Delhi's communications officer told the Comet to go ahead with its message, there was no reply. At the same time that afternoon, workers in paddy fields near Jagalgori, some 20 miles (32km) west of Calcutta, heard a loud bang during a thunderstorm and saw a blaze of fire in the sky. Pieces of burning aircraft wreckage and bodies started to rain down around them. The wreckage was spotted the next morning from a BOAC York freighter, piloted by no less than Captain Harry Foote. He must have had mixed feelings.

But at the inquiry, again the pilot was blamed.

Another accident followed. Taking off from Rome's Ciampino Airport on 10 January 1954, a BOAC Comet climbed to 27,000ft over Italy's west coast. The captain called a BOAC Argonaut aircraft, which had taken off from Rome just ten minutes before, possibly to ask about a weather report he had just transmitted. 'George How Jig from George Yoke Peter did you get my...' At this point, the transmission broke off.

At the same moment, fishermen at sea off the coast of Elba heard a loud series of explosions. Looking up from their nets, they saw pieces of the Comet, some on fire, fluttering down into the sea.

An extensive search by Royal Navy ships began and bodies were recovered, many of which had strange injuries: they had been stripped half naked and several had fractured skulls. An inquiry began and BOAC announced the suspension of all Comet flights.

All kinds of speculation arose: was there something unlucky about Rome? Could the aircraft have been broken up by jet-stream winds? Had a jet turbine wheel exploded? Could a bomb have been placed on board? One Sydney newspaper even smuggled a dummy bomb aboard a conventional airliner. 'It Can Happen Here,' the front page trumpeted ('You Wouldn't Get Away with That Now,'[1] we might reply).

---

1 This was the Sydney *Truth*, now part of the Murdoch News Corp empire.

No one had any idea what had gone wrong.

Some shielding was put in place between the turbine wheels and the fuel systems, and nine weeks later Comet flights resumed.

The investigation dragged on, but before long there was yet another Comet crash, on 8 April 1954. It was a BOAC aircraft, G-ALYY ('Yolk Yolk'), operating a South African Airways charter.

Once again, the aircraft had taken off from Rome. It had reached much the same altitude as 'Yolk Peter', the previous Comet, and the two airframes had a similar number of flying hours on them. This time the Comet exploded off the coast of Sicily near the volcanic island of Stromboli and fell into the sea. This was all too much, and the Comet's British certificate of airworthiness was withdrawn on 24 April 1954. RAE Farnborough took charge of the investigations, and these were to become the model for the world-standard aircraft crash investigations that we have now.

If you are wondering why it took so long to ground the Comet, it must be remembered that air crashes were a far more frequent occurrence in those early years. For example, 40 pilots had been hired by the US Postal Department in 1918, half of them were dead after the first year and within ten years, 31 had died in air crashes. If the 1929 US passenger death rate applied today, 7,000 would die every year in the United States alone. It was no better after the war: piston-engined propeller-driven airliners were crashing regularly and more passengers were dying because those planes were larger than the first-generation Comets. In 1950, an Avro 689 Tudor V stalled and crashed in Wales, killing 83 people. At the time, it was the worst air disaster in history. Douglas DC-6s crashed three times in 1952, with 107 deaths total, three times in 1953 with 123 killed, and three times in 1954, with 63 dead. Convair CV-240s crashed three times in 1953 – 88 dead in all. And only four days *after* the Comet crash over Elba, another Douglas DC-6 had crashed and burned while landing at Rome, killing all 16 people on board.

The wreckage of 'Yolk Peter' had been laboriously recovered from the seabed off Elba (Comet 'Yolk' was in waters that were too deep) and pieced together in a hangar at Farnborough.

Were the crashes something to do with Rome airport? In particular, the way fuel bowsers pumped fuel into the wing tanks under pressure. And so, could the wings have been buckled by high-pressure fuel filling? Suspicion first fell on an Italian Viberti fuel bowser lorry that had attended 'Yolk Peter', and it was brought from Italy. However, it was found that it pumped at a lower flow rate and half the pressure of the British Dorset fuel bowsers. That was not the cause.

Did the injuries to the passengers reveal any clues? Examination of the bodies revealed fractured skulls and lungs that had exploded from sudden depressurisation. What exactly had happened inside the cabin? A scale model of the aircraft was built at Farnborough, fitted with small seats containing doll-like figurines and pressurised in the same way as the real aircraft cabin. It was then explosively blown open. A high-speed camera recorded the result. It was shocking in its graphic detail: a whirling storm of wrenched-off seats, flying bodies and shredded aluminium. And it all happened so quickly.

At 0.003 seconds after the test explosion, the backs of the passenger seats slammed forward, smashed down by a gale of air rushing out. The cabin began to break up. At 0.009 seconds, seats were flying towards the primary hole. At 0.25 seconds, one dummy followed out feet first. At 0.6 seconds, more dummies flew out of the aircraft. One smashed its head against the roof. At 1.0 second, it was all over. The wreckage was like a cloud of confetti, some of it burning.

There was a turbine wheel missing from one of the Elba crash engines. Had it burst at high rpm and exploded through the pressurised cabin, causing it to pop like a balloon? No. When the port wing was examined, it was found that the missing turbine disc had cut through it like a circular saw, exiting through the upper surface when the upside-down wing had hit the sea. The turbine wheel, still spinning, had just kept going to the bottom of the sea.

Did the jet exhaust, bouncing off the runway when taking off and playing on the cabin sides, somehow weaken them with heat? Experimentation found that this indeed was the case, and the factory obviously knew about this, as repair patches had been put in place. This certainly was a weakness, and later models of the Comet had the jet efflux angled slightly away from the cabin sides. But this was not the cause.

Back at Farnborough, a Comet fuselage had been placed in a water tank and stressed with repeated pressure changes equivalent to thousands of hours of flying time. On 24 June, this cabin suddenly burst. Sir Geoffrey de Havilland and designer Ronald Bishop rushed to the tank. To their horror there was a gash on the port side of the cabin 8ft (2.5m) long and 3ft (1m) deep. There was also a tell-tale discoloration of the aluminium at a rivet hole at the bottom rear corner of the escape hatch: metal fatigue. There was another crack, over an inch long, on the top corner of a window on the port side. This was also a fatigue crack, and the beginning of another major failure point.

Metal fatigue is what makes paper clips snap when repeatedly bent to and fro: it is the weakening of metal caused by cyclic loading.

When the top of the cabin of the Elba Comet was recovered, the cause of the crash was found and, at last, (they said) the Comet riddle was solved.

The two square black ADF windows had reinforcing frames around them. The rivets holding them to the skin were placed closer to the edge than the designer had specified. These frames had only been rivetted, not glued as well as the design specification called for. These rivet holes were not drilled but punched, and the irregularities in such a hole can be the beginning of a fatigue crack. Then, more manufacturing faults were found. When the Comet had been built, the forward ADF reinforcing frame had a small crack accidentally made in it. The workman had drilled a small hole in the end of the crack as a way of stopping it. This same crack-and-hole-drilling technique in the ADF frame had been spotted in 'YU', another Comet. Although acceptable in some workshop situations, this mistake could have been enough to take down the lightly built Comets. These cracks had been the fatal starting points for metal fatigue.

'You may know nothing at all about it,' said Sir Lionel Heald, the Crown representative, 'until a disastrous fracture suddenly occurs without warning'.

What had happened, it was thought, was that cracks had started around the rear ADF window, which caused surrounding panels to blow out and flip back, hitting the rear elevators. The rear of the fuselage broke away, then the wings and cockpit ripped off. It would all have happened in less than a second. It was like the pop of a balloon.

The investigation had cost over £2 million (around £19 billion today) and now there was a full public inquiry. This was conducted in a transparent way that would not have happened in the Soviet Union or the repressive regimes of today. The information gained at the cost of so many lives and treasure was freely shared with the rest of the world of aviation, including Britain's competitors. Lord Brabazon said during the inquiry:

> You know and I know the cause of this accident. It is due to the
> adventurous pioneering spirit of our race. It has been like that in

the past, it is like that in the present, and I hope it will be in the future. In this inquiry there is nothing to be ashamed of; much more to be proud of. Here was a great imaginative project, to build a machine with twice the speed and twice the height of any existing machine in the world. We all went into it with our eyes wide open. We were conscious of the dangers that were lurking in the unknown. We did not know what fate was going to hold out for us in the future. Of course, we gave hostages to fate, but I cannot believe that this Court, or our country, will censure us because we ventured. You would not have the aeronautical people in this country trail behind the world in craven fear lest they be censured in such a Court as this for trying to lead the world. Everything within the realm of human knowledge and wisdom was put into this machine.

When we gave this Certificate of Airworthiness to these machines, they were airworthy. True, they deteriorated in a way no one on earth at the time could foretell, and they deteriorated, so I am led to understand, by a slowly developing molecular metallurgical fault. It is metallurgy, not aeronautics, that is in the dock.... My Lords, I have now had fifty years connected with aviation, and if I may say something about it, I would like to say this. That in every step in progress we have paid for it in blood and treasure, and God knows that in this case we have paid in full. Finally, I do hope that the threat of having to face an inquiry such as this, with all its publicity, if anything goes wrong, will not stop adventurous spirits pioneering in the future.

De Havilland was not found criminally liable for the Comet's disintegration. The Court of Inquiry found it had worked with the best accepted practice. As Sir Arnold Hall put it, 'They built the Comet around the most advanced engineering knowledge available and could certainly not be held to blame.'

Not everyone was happy about the decision, but Geoffrey de Havilland decided to go ahead with an improved version of the

Comet: 'To many of us it was the shattering blow of sudden and complete failure following on notable achievement that was the hardest to bear… We felt that it must be possible to seek out and remedy the unknown failure in a short time… The idea of giving up on the Comet was never seriously contemplated. It had just got to be made right.'

The *New York Herald Tribune* concluded, 'Full marks to Britain for its brutally honest and frank inquiry into the Comet.'

That was not quite the end of the story. In 1956, another piece of the Elba Comet wreckage was trawled up 15 miles (24km) south of where the original wreckage had been found. This piece was from the starboard side of the cabin, just above the three front windows and just below the ADF windows. Careful examination at Farnborough then suggested that the primary failure was probably near to this area rather than at the rear ADF window on the roof of the cabin, as had been previously thought. It was still metal fatigue. These findings, however, were kept secret until the details were published in 2015.

In 1952, Trans World Airlines (TWA), owned by the secretive and increasingly eccentric Howard Hughes, had considered buying the Comet. In later years, with the benefit of 20:20 hindsight, TWA's vice-president, Robert Rummel, commented:

I considered the Comet program to be a superb pioneering venture that quite obviously required pressing the state of the art of airplane design in nearly all significant technology areas to achieve the barest minimum acceptable overall efficiency. In the earlier models this produced marginal structure, minimal operating weights, and borderline performance. For example, the thin fuselage skin of the Comet had been stretched during manufacturing to increase strength at the expense of ductility; every pound of empty weight was critical re payload or range; higher-thrust engines were clearly needed. The limited range, sluggish take-off at high rotation angles, and the ability to stop after landing on slippery runways were also important concerns. Except for the lack of reverse thrust, I did not think any one of the marginal conditions ruled the airplane out, but the combination of them gave me serious pause.

In the end, the Comet 1 was just too lightly built. This was because the Ghost jet engines were a first-generation design with low thrust and high fuel consumption. They were insufficiently powerful to propel the heavier, stronger, safer airframe that the Comet should have had, and yet still provide a profitable payload. The new powered controls were also probably too light and too dead, making it too easy to over-rotate and failing to provide pilots with enough feedback of an imminent stall. Metal fatigue was too little understood. Modern aircraft now use thicker skins with internal cross webbing that inhibit the spread of cracks.[2] Windows are smaller. Controls provide more feedback. Jet engines are more powerful.

There was a curious prophesy of the Comet 1 disasters. Five years *before* the accidents, the British author Nevil Shute had published a novel, *No Highway*, in which metal fatigue, after a number of flight cycles, brings down a fictional new airliner. It is named the *Rutland Reindeer*, and you might notice an odd parallel: one of Santa Claus' reindeers is named Comet. The aircraft company's owner seems to be a portrait of Geoffrey de Havilland. The anti-hero appears to be a portrait of Alan Griffith. Also, Shute

---

2 These are termed 'rip-stop doublers' or crack arresters. They are extra strips of aluminium placed concentrically within the fuselage skin, which will stop the progress of a crack.

himself had worked for the company as an aeronautical engineer.[3] Did he suspect something? This is one of the mysteries of aviation that will never be solved.

Only 21 Comet 1s and 1As were built, and all were either scrapped or used in stress testing. Before the Comet 1 disasters, a second series was already well under way. The Comet 2 had oval windows, larger wings, a larger fuel capacity and a heavier-gauge skin. But all outstanding orders for the Comet 2 were cancelled by airline customers.

Only one flying example of a Comet 3 was built. This was a Comet 2 lengthened 15ft 5in (4.70m) and powered by Avon M502 engines developing 10,000lbf (44kN), which first flew on 19 July 1954. This led to the definitive and most successful Comet: the Series 4. This is the design that the Comet should have been all along. This aircraft initiated the transatlantic service, and deliveries to BOAC began on 30 September 1958 with two 48-seat aircraft. It was a commercial success, with 74 built.

The Comet 4 was 18ft 6in (5.64m) longer than the Comet 1 and seated 74 to 81 passengers compared to the first Comet's 36 to 44. It had a longer range, a higher cruising speed and a higher maximum take-off weight than the Comet 1, and these performance improvements were largely due to the new Rolls-Royce Avon axial-flow engines, which had twice the thrust of the Comet 1's Ghosts.

The Comet was not alone in having this type of failure. In 1988, on a flight between Honolulu and Hilo, the roof of a Boeing 737 blew off in an explosive decompression. A flight attendant was sucked out of

---

3 A brilliant writer with a technical bent, Shute also worked with Barnes Wallis. *On the Beach* and *Trustee from the Toolroom* – recommended.

the hole and 65 passengers and crew were hurt. This was blamed on faulty lap joints in the aluminium skin of the fuselage.

The de Havilland Comet was a brave effort by a nation exhausted by war. At first it seemed as though there was to be a new Elizabethan age, with a beautiful young queen on the throne, Mount Everest climbed and a British jet airliner racing through the skies.

Somehow that all made the Comet's downfall harder to take.

In the end, the Comet jet airliner was one of the most spectacular and futuristic aircraft ever built. Seventy years on, jet airliners still look much the same. Its downfall was just as spectacular.

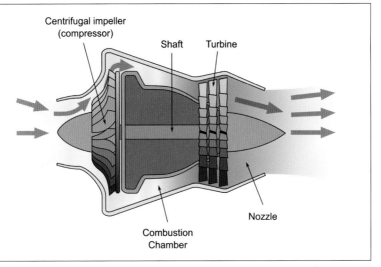

Centrifugal impeller
(compressor)

Shaft   Turbine

Nozzle

Combustion
Chamber

How Whittle's jet engine worked. Air is drawn in, compressed, fuel is burned to rotate the turbine and the exhaust provides the thrust. (Original design: Emoscopes (talk), Vectorization: Tachymètre, CC BY 2.5, https://creativecommons.org/licenses/by/2.5, via Wikimedia Commons)

The piston aero engine had become horribly complicated: a 54-litre Bristol Centaurus. (Key Archive)

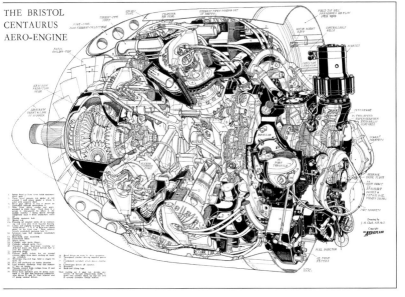

Frank Whittle loved the manoeuvrable Armstrong Whitworth Siskin. (Key Archive)

Whittle (centre) works in the test room on his first jet engine. (TriFocal Communications/ Stocktrek Images/ Getty Images)

Whittle's engine installed in the Gloster E.28/39. (Key Archive)

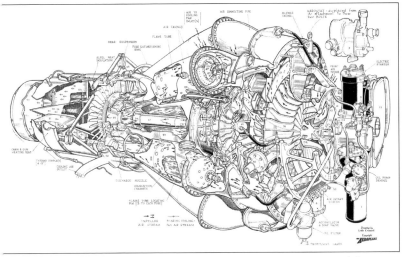

The Rolls-Royce Derwent improved on Whittle's ideas. This cutaway shows airflow through the engine. (Key Archive)

The US was quick to exploit Whittle's design: the Allison J33-A-15 turbojet engine. (Hunini, CC BY-SA 4.0, https://creativecommons.org/licenses/by-sa/4.0, via Wikimedia Commons)

Gloster Meteors were a pleasure to fly. (DC, Key Archive)

The first jet-powered aircraft: the Heinkel He 178. (Key Archive)

Me 262 captured intact on the airfield at Giebelstat. (Key Archive)

The Me 262 was the world's first operational jet-powered fighter aircraft. (USAF)

The swept-wing MiG-15 used a copy of the Rolls-Royce Nene engine. (Key Archives)

Above: The MiG-17 could approach Mach 1, the speed of sound. (Key Archives)

Right: Surprisingly, the first American jet fighter, the Lockheed P-80 Shooting Star, was powered by a British engine. (Key Archives)

The swept-wing North American F-86 Sabre could take on and beat the MiG-15s. (Key Archives)

This test Lancastrian VM703 was fitted with Merlin piston engines, Ghost jet engines and rockets. (Key Archives)

The world's first jet engine to enter airline service: the de Havilland Ghost. (Key Archives)

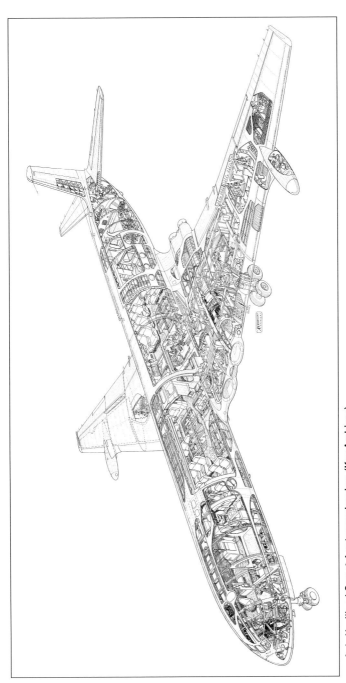

A de Havilland Comet 4 cutaway drawing. (Key Archives)

We were always told to wave! (Key Archives)

The amazingly prescient supersonic Miles M.52. (Key Archives)

The troubled Howard Hughes on the flight deck of his *Spruce Goose*. (Key Archives)

The de Havilland DH 108 Swallow: a killer of test pilots. (Key Archives)

A river of power: the Rolls-Royce Avon. (Key Archives)

Warfare old and new: camels with the English Electric Canberra in Aden, 1957. (Key Archives)

Above: The enormously successful Canberra, with 1,352 built. (Key Archive)

Left: The Canberra's twin Rolls-Royce Avon engines enabled a high rate of climb. (Key Archives)

Above: The Lockheed U-2 ran on cigarette lighter fuel: 'That bird was a giant flying Zippo.' (USAF)

Right: Pacific Southwest Airlines flight attendants in their microskirts and boots uniform. (San Diego Air and Space Museum)

The Boeing 707-436 Jet Transport

Four 17,500 lb. s.t. Rolls-Royce Conway 508 By-pass Turbojet Engines

© Temple Press Limited, 1960

Drawing by B. Hatton

This Boeing 707 used Rolls-Royce Conway engines. (Key Archives)

The world's first gas turbine car: JET 1. (Jbattersby, Public domain, via Wikimedia Commons)

An Avro Vulcan testing a fifth, bifurcated inlet Olympus engine. (Key Archives)

Stratosaurus Rex. The Boeing B-52 bomber mounted no fewer than eight twin-spool JT57 engines. (Key Archives)

The luxurious interior of the Tupolev Tu-104. (Key Archives)

A Russian stewardess prepares caviar during a Tupolev Tu-104 flight. (Key Archives)

Above left: Jackie Kennedy made the Boeing 707 fashionable. (National Archives)

Above right: Zero to 1,800mph in three seconds: the ramjet-powered Bristol Bloodhound. (Key Archives)

Vith two Bristol Olympus engines, the TSR.2 had immense speed. Even the Lightning hase aircraft struggled to keep up with it. (Key Archives)

THE ARMSTRONG SIDDELEY MAMBA

Airscrew-Driving Turbo-Jet

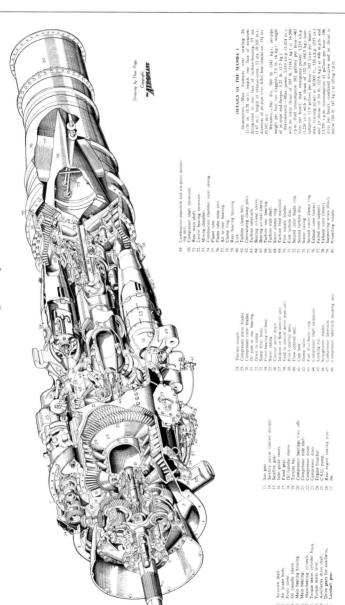

Drawing by Thos. Page.

Copyright *AEROPLANE*

The jet engine could also drive a propeller (Key Archives)

1. Airscrew shaft.
2. Air intake body.
3. Front cover.
4. Oil transfer sleeve.
5. Main bearing housing.
6. Main bearing.
7. Main-bearing oil-seals.
8. Torque motor cylinder block.
9. Torque motor lever.
10. Auxiliaries drive shaft.
11. Drive gears for auxiliaries.
12. Layshaft gear.

13. Sun gear.
14. Satellite carrier (shown divided).
15. Inlet guide vanes.
16. Fixed gear.
17. Oil transfer sleeve.
18. Torsion bar.
19. Torsion bar.
20. Compressor bearings (two off).
21. Compressor stub shaft.
22. Compressor drum.
23. Compressor discs.
24. Engine breather.
25. C.S.U.
26. Rear engine cooling pipe.
27. Jet.

28. Thermo couple.
29. Compressor stator blades.
30. Compressor rotor blades.
31. Oil pipe to rear bearing.
32. Drain to sump.
33. Sump drain oil-feed.
34. Front-bearing oil-feed.
35. Stator casing.
36. Centrifugal drain.
37. Insulator to flow control unit.
38. Feed to control servo pipe unit.
39. Pilot's control lever.
40. Flow control unit.
41. Cam box.
42. Dump valve.
43. Fuel distributor ring.
44. Compressor shaft extension.
45. Locking nut.
46. Igniter.
47. Compressor labyrinth.
48. Compressor labyrinth housing unit.

49. Combustion manifold and six-point mounting unit.
50. Compressor shaft. extension.
51. Rear main shaft.
52. Centre bearing extension.
53. Mixing chamber.
54. Combustion chamber outer casing.
55. Flame tube.
56. Flame tube nose unit.
57. Air to rear bearing.
58. Clamp ring.
59. Rear bearing housing.
60. Link.
61. Turbine clamp bolt.
62. Centralizing clamp plate.
63. Turbine labyrinth.
64. Bearing oil-seal sleeve.
65. Bearing air-seal sleeve.
66. Turbine rear bearing.
67. Turbine stub shaft.
68. Turbine clamp ring.
69. Turbine feed manifold.
70. First nozzle blades.
71. First turbine blades.
72. First turbine disc.
73. Second inner blade ring.
74. Second turbine disc.
75. Stator casing.
76. Second stator clamp ring.
77. Exhaust cone (outer).
78. Fixed cone support.
79. Supporting struts (four).
80. Propelling nozzle.

DETAILS OF THE MAMBA 1

Dimensions.—Max. diameter over cowling 29,11/16 in. (0.79 m.); length, rear face of airscrew-fixing-cone to rear face of turbine-housing 57 ins. (1.47 m.); length of exhaust-cone, 12 ins. (0.303 m.); diameter of jet-pipe over Altol heat-insulation, 11¾ ins. (0.305 m.).

Weights.—Net dry, 760 lb. (343 kg.); jet-pipe weight per foot run (lagged) 7.9 lb. (4 kg.); weight of jet-pipe end-flanges, 3.25 lb. (1.7 kg.)

Performance.—Max. power, 1,010 s.h.p. (1,024 c.v.) with jet static thrust of 307 lb. (138.5 kg.) at 14,500 r.p.m. (fuel consumption 102 gallons per hour—465 litres per hour); max. emergency power, 1,210 s.h.p. (1,226 c.v.), with jet thrust of 152 lb. (68.5 kg.) (consumption 11.9 gallons per hour—507 litres per hour); max. cruising power at 30,000 ft., 550 s.h.p. (557 c.v.) and jet thrust of 50 lb. (22.6 kg.) at 400 m.p.h. and 13,750 r.p.m. (consumption 41 gallons per hour—186 litres per hour); combined airscrew and jet thrust is below 100 lb. (45 kg.) at idling r.p.m.

Chapter 7

The Speed of Sound

After the war, aircraft makers turned their energies towards a less visible enemy: the sound barrier.

The term is an anomaly; to the fascinated public, it suggested an invisible wall somehow to be penetrated by the sharp nose of a jet aircraft. Instead, as we have seen with the propeller-less Spitfire, the barrier is an increase in aerodynamic drag experienced by an aircraft as it approaches the speed of sound. This increased drag makes it difficult to exceed the speed of sound unless the aircraft is specially designed to overcome the drag effects.

The speed of sound on a dry, warm day at ground level is about 767mph, or 343 metres per second (1,234km/h). Light is a million times faster than sound. In naval engagements before 1850 it was normal to see the flash of the enemy gun, then hear the boom a few seconds later; only after that would the subsonic shot come sailing through the air. Sharp-eyed sailors learned to duck. Was this the origin of 'Flash, bang, wallop.'? After the 1850s, shells managed supersonic speeds.

The temperature and the medium in which the sound travels affect the speed: the speed of sound in cold thin air above 35,000ft is 660mph (1,063km/h). And the speed of sound in diamonds is 27,000mph! (12,000 metres per second).

Because there is no absolute speed for sound (as there is for light), aeronautical engineers had to come up with a relative measure, so the speed of sound in air is called Mach 1.0 (after the Austrian Ernst

Mach). So, an aircraft flying Mach 1.0 at sea level is doing around 767mph, (1,234km/h, 666 knots), but a plane flying Mach 1.0 at 35,000ft is flying at around 660mph (1,063km/h, 589 knots) etc. An aeroplane flying at Mach 2.0 is flying twice as fast as the speed of sound. Speeds below Mach 1 are called subsonic, between Mach 0.8 and 1.2 transonic, and above Mach 1.2 supersonic. Above Mach 5 is called hypersonic.

There is nothing particularly unnatural about exceeding the speed of sound. Meteors cracking through the Earth's atmosphere do it. Bullets, shells and rockets do it. Even the crack of a bullwhip is caused by the tip breaking the sound barrier. Dinosaurs with long tails such as the Brontosaurus, Apatosaurus and Diplodocus may have been able to flick their tails at supersonic speeds, with what must have been a terrifying crack echoing through the tropical forests. And a London bus could go supersonic if it had enough thrust.

I remember hearing my first sonic boom on a French beach in the early 1960s, probably made by a Dassault Mirage jet fighter. The sound was a thunderous double bang in the sky, which filled all four corners of that long-ago morning.

Just as a ship will make bow and stern waves, so too will an aircraft make a series of pressure waves in front of it. These waves travel at the speed of sound and as the speed of the aircraft increases, the waves are compressed together because they cannot get out of each other's way fast enough. When that happens, they merge into a single shock wave, which travels at the speed of sound. The double bang is the sound of the shock waves at the front and rear of the aircraft.

To the little boy on the shore, that sonic double bang was one event, but, in fact, the aircraft was laying down a continuous boom all along its course along the coast, like the rolling-out of a carpet. The supersonic pilot hears nothing: 'You don't actually hear anything on board. All we see is the pressure wave moving down the aeroplane – it gives an indication on the instruments. And that's what we see around Mach 1. But we don't hear the sonic boom or anything like that. That's rather like the wake of a ship – it's behind us.'[1]

During the war, there had been records claimed by the Messerschmitt Me 262 at 624mph (1,004km/h) and the rocket-powered Messerschmitt Me 163A Komet at 702mph (1,130km/h) but these were under wartime conditions. No evidence of such speeds exists in any of the research from that period, which was captured and exhaustively studied by the Allied forces. But the unmanned V-2 rockets that plunged onto London had routinely exceeded Mach 4 on re-entry, producing a sound like thunder. Had the Germans also broken the sound barrier in a manned aircraft? Captain E. M. Brown carried out transonic flight-testing of the Me 262 at Farnborough in the summer of 1945 and concluded that the limiting Mach number was 0.86: 'Beyond this control would be lost as aerodynamic forces exceed the strength of the pilot.'[2]

The Bachem Ba 349 Natter (grass snake in English) was a rocket-propelled surface-to-air missile built by Bachem Werke GmbH to deal with the hordes of Allied bombers flying over Germany night and day. The rocket was built of wood held together with nails, but the motor itself was highly effective: the Natter was expected to exceed 620mph (1,000km/h). The only problem was that no automatic homing system could be made to work as well as a human pilot.[3] So it was designed with a cockpit and the means for a

---

1 http://news.bbc.co.uk/1/hi/talking_point/3207470.stm

2 Letter to *The Telegraph* (February 2013)

3 'Project Pigeon' used pigeons as pilots in US glider-bombs. They were trained to peck at a screen showing an image of the target, steering the missile. Inventor B. F. Skinner complained, 'Our problem was no one would take us seriously.' The programme was cancelled on 8 October 1944.

pilot to exit after firing rockets at the enemy. The idea is that after the attack the Natter would level off and the pilot would bail out.

A young Luftwaffe test pilot, Lothar Sieber, volunteered for the first (and last) manned flight. He took off vertically, but the 3g forces probably caused him to haul back on the control column, because the Natter climbed into an inverted curve; the cockpit canopy was seen to fall off and the Natter disappeared into cloud. It crashed some distance away. Poor Sieber was killed, but he was certainly the first man to take off vertically from the ground under pure rocket power, 16 years before Yuri Gagarin's Vostok 1 flight.

The British Air Ministry ran a secret project during the war that is little known about today: an attempt on the speed of sound itself. Not only that, but it also aimed far higher – for a speed of 1,000mph (1,600km/h), a speed hitherto undreamed of. And it all came about because of a misunderstood radio transmission.

During the 1930s, the accepted wisdom was that manned flight above the speed of sound was impossible due to the problems of compressibility (the increased drag due to shock waves). Towards the end of the war though, three supersonic programmes were started: one British, one German and one American. As Bill Gunston commented: 'It is typical of the way we do things in Britain that we started before our rivals, had a distinctly superior design and then, at the eleventh hour, cancelled it for political reasons.'[4]

However, in 1942, the British Air Ministry was panicked by an intercepted communication reporting that a German aircraft, either a Messerschmitt Me 163A or a Me 262, had just exceeded 1,000mph (1,600km/h). This was shocking indeed, considering that most contemporary piston-engined fighters struggled to exceed 400mph

---

4 Gunston, Bill, *Faster Than Sound*, Patrick Stephens, Sparkford (1992)

(644km/h). Memos flew like confetti and the ministry issued Air Ministry Specification E.24/43 in autumn 1943. This was probably the shortest and most ambitious requirement ever issued by that department. It called for an aircraft capable of taking off from a normal airfield, flying at over 1,000mph (1,600 km/h) in level flight, then for the ability to climb to 36,000ft in 1 minute 30 seconds. It would necessitate exploring a whole new realm of aeronautics.

The choice of aircraft company to build this projectile was another surprise. The Miles Aircraft company, run by brothers Fred and George Miles, had quietly been making rather pedestrian single-engined aircraft such as the 140mph (230km/h) Miles Magister and 220mph (350km/h) Miles Master, used as trainers for the RAF. However, Sir Stafford Cripps owed the company a favour and liked Fred Miles. Miles was married to Maxine 'Blossom' Miles, who had been presented as a debutante with Barbara Cartland and Lady Diana Cooper. This talented woman became an aviation engineer and draughtswoman, designing the Miles Hawk G-ACIZ aircraft, which Gabrielle Patterson, Britain's first woman flying instructor, flew in the King's Cup air race in 1934.

Fred Miles was a remarkably far-sighted designer: he had been manufacturing a new ballpoint pen for the RAF. This had been invented by the Hungarian-Argentinian László Bíró. It was named the Biro. So Miles Aircraft might have been tiny, but its designers were talented, and they worked fast. As George Miles said, 'Of course, if we'd been big, we'd have been far slower.'

As there was no relevant aircraft technology for 1,000mph (1,600km/h) speeds, Fred Miles turned to the field of ballistics and decided that, as his aircraft would have to travel at the speed of a bullet, it may as well look like a bullet. The design had a sharp conical nose and sharp leading edges on the wings, as it was known from artillery shell research that round noses were unstable above the speed of sound. The model prototype's wing leading edges were so sharp that the mechanics kept cutting themselves, so much so that they christened the aircraft 'The Gillette'. It had short, clipped wings to keep them out of the cone of the shock wave, and these had a thin biconvex section for the least drag.

And so, the Miles M.52 was born. It was to use one of Frank Whittle's W.2/700 turbojet engines to power it up to 1,000mph (1,600km/h). The engine would use a reheat jet pipe, now known as an afterburner. Extra fuel was to be burned in the tailpipe to avoid overheating the turbine blades, making use of unused oxygen present in the exhaust. To supply more air to the afterburner a thrust augmenter fan powered by the turbine was to be fitted behind the engine to draw air *around* the engine via ducts: an early example of a turbofan.

Whittle knew that it was more energy efficient to accelerate a large amount of air by a small amount than it was to accelerate a small amount of air by a large amount, even though the thrust was the same. This is one reason why modern high-bypass turbofan engines are more efficient than simple jet engines at subsonic speeds. The turbofan or bypass engine now powers most modern aircraft and is likely to do so for the foreseeable future. It can be seen as a hybrid of the turboprop and the turbojet engine. The thrust from a bypass engine is derived from the mass air flow from the 'fan' plus the mass air flow from the core engine and can be exhausted separately or mixed prior to entering the jet pipe.

Whittle had applied for patents for this idea back in June 1936.[5] When Whittle's drawings of the scheme were given to General Electric in 1942, someone took notice. No acknowledgement of Whittle's idea is mentioned in the company history book, *Seven Decades of Progress*, even though precisely the same scheme was adopted in the company's first production turbofan in 1961. We will return to the turbofan when it became mainstream thinking.

Whittle's improvements would have resulted in 1,620lb of extra thrust for the Miles M.52 at 36,000ft and 500mph (800km/h).

5 UK Patent 471368.

Further gains in power were expected above the speed of sound. A shock cone in the nose would slow down the air to subsonic speeds, as Miles and his designers knew the engine could not digest a supersonic flow of air. The whole design was extremely advanced and betrays the fact that Miles and his designers had learned a great deal about supersonic aerodynamics.

The cleverest feature of the Miles M.52 was one that was to be universally adopted by most supersonic aircraft: the variable-incidence, or 'flying tail'. At the subsonic speeds reached by fighters during dives, their conventional control surfaces revealed a cloven hoof: they became either ineffective or even reversed in function. Shockwaves built up at the hinges of the old-fashioned hinged tailplanes or rudders, moving the centre of pressure rearwards and making the controls so heavy that the pilot could not recover from a dive (much was made of this in David Lean's 1952 film, *The Sound Barrier*, when a counter-intuitive control movement saved the day).

Instead, on the Miles M.52, the whole tailplane rotated in the manner of a balanced rudder on a racing sailing yacht, rather than having a hinged panel hanging at the trailing edge of the tailplane. Also, the controls were powered so that the pilot's strength could not be overcome. This feature had been tried on some Spitfires used for high-speed tests. In contrast, all early jet aeroplanes of the time had rounded noses, thick wings, manually powered controls and hinged elevators so they couldn't safely approach trans-sonic or supersonic speeds.

The Miles M.52 design was a tour de force of supersonic understanding and it was extraordinarily prescient. Aeronautical historian Derek Wood observed that the M.52 contained 'all the ingredients of a high-performance aircraft of the late fifties and even some of the early sixties'.[6]

---

6 Wood, Derek, *Project Cancelled*, The Bobbs-Merrill Company, Inc., Indianapolis (1975)

By 1944, the Miles M.52 design was ready, and the company was told to get on with the building of three prototypes.

Then the dead hand of the bureaucrat was laid over the designer's throat.

At some point, it had been realised that the original radio transmission had been mistranslated: it was 1,000 *kilometres* per hour, not miles per hour. It is unclear whether that had any bearing on the ensuing decisions.

The Air Ministry had signed up to an agreement with the United States to exchange supersonic research. This was understood by the British to be a reciprocal arrangement. The Bell Aircraft company was given access to the drawings and research on the M.52, and Miles' chief aerodynamicist, Dennis Bancroft, stated that Bell Aircraft personnel visited Miles later in 1944 and were given access to all the drawings and research on the M.52.

Unbeknown to Miles, Bell had already started building a rocket-propelled supersonic design, which had a conventional tail.[7] As a result, they were struggling with pitch control at supersonic speeds. A flying tail like that of the Miles M.52 was built onto the Bell prototype and the test pilot, Chuck Yeager, reported that it now worked.

The Bell X-1 achieved a speed of nearly 1,000mph (1,600 km/h; 870 knots) in 1948, and Yeager became the first pilot to exceed the speed of sound. It is notable that it had straight wings.

It must be remembered that the X-1 was rocket propelled rather than jet propelled, and it had only 4 minutes 40 seconds of powered flight. Jet fighters needed rather longer than that to do their job. However, thanks to the Miles' design, all future supersonic aircraft would have either an all-moving tailplane or a delta wing.

Six months after the end of the war in Europe, Fred Miles was told to stop work on the M.52. Rather like Frank Whittle, he was

---

7 Miller, Jay, *The X-planes: X-1 to X-45*, Midland Publishing, Birmingham (2001)

stunned by the news: 'I did not know what to say or think when this extraordinary decision was sprung upon me, without warning of any kind. At our last official design meeting all members, including the Ministry and Power Jets' representatives, had been cheerful and optimistic.'

Unfortunately, in late 1945, a great deal of the captured German research appeared to concentrate on swept wings. Someone at the Ministry of Supply must have thought that the Miles M.52 was mistaken in having straight wings. Gunston comments:

> I still cannot comprehend how it was that none of the aerodynamicists explained to them that sweepback is useful only in postponing the drag rise in the transonic regime. If you are going to fly at highly supersonic speeds then you forget about 'the Barrier' and settle down in the regime on the other side and sweepback is meaningless... Ministry department heads were asked for their views on the programme, and their replies showed almost total ignorance of the difference between a transonic wing and a supersonic one.[8]

Once again, the British Achilles' heel had been cruelly exposed: the lack of scientific knowledge among the nation's classically educated administrators. The reasons offered were to do with the budgetary constraints on the Attlee government and worries about safety.

In 1947, Miles Aircraft Ltd entered receivership. The cancellation set back British progress in the field of supersonic design by ten years.

It is quite astonishing that, back in 1933, when he was flying a 100mph (160km/h) biplane, Frank Whittle had understood what

---

8 Gunston, *Faster Than Sound*

was required to fly an aircraft faster than sound. A propulsion system that dispensed with the propeller was needed, and so he invented the turbojet. He wanted to go even faster, so he also invented the afterburner to increase the jet velocity to allow supersonic flight. The afterburner burned additional fuel in the jet pipe downstream of the turbine. With no blades to worry about, the temperature could be raised to white heat, or high as 1,750°C. Indeed, this man was a genius.

The RAF High Speed Flight was a small group of pilots that came together to win the Schneider Trophy for racing seaplanes in the last years of the 1920s. They flew the ancestor of the Spitfire, the Supermarine S6B seaplane, which won the trophy for Britain in perpetuity. That aircraft was powered by the Rolls-Royce 'R' racing engine, a piston engine that at one point held the world land, air and water speed records simultaneously, a feat unlikely ever to be repeated.

Rather oddly, in the light of the supersonic project the Air Ministry had just cancelled, the flight was reformed by the RAF in 1946 to improve on its official world airspeed record of 606mph (975km/h), set the year before by Hugh Wilson in a Gloster Meteor, the first official post-war record.

It was a publicity stunt. The RAF's High-Speed Flight was equipped with two Meteor IVs that were specially prepared for the speed record attempts. The Rolls-Royce Derwent engines had a mild increase in power, and an aluminium cockpit hood was fitted as the normal Perspex hood was beginning to melt with the air friction at over 600mph (970km/h). The flight barely improved on Wilson's 606mph (975km/h); the highest speed achieved being 616 mph (991km/h). It was still a new world record, and perhaps the foreign buyers of the Meteor were impressed. But as supersonic research came under the Official Secrets Act, Fred Miles was unable to protest to the newspapers about the cancellation of the 1,000mph (1,600km/h) M.52, and so the public didn't know that a British supersonic aircraft had so nearly been built. Nor did they know of its inexplicable termination. Some might have thought:

'For of all sad words of tongue or pen,
The saddest are these: "It might have been!"'[9] '

One solution to the problem of a conventional tail on a supersonic aircraft was to cut it off altogether. And so that's what Geoffrey de Havilland did. An early design for his Comet airliner had swept wings and no tail, like the German Messerschmitt Me 163 Komet rocket-powered point-defence interceptor. To test this delta[10] wing, as it was called, and to investigate stability and control problems of this kind of wing an experimental aircraft was built, the de Havilland DH 108. The designer, John 'Jack' Frost married the front fuselage of the new de Havilland Vampire jet fighter to a swept wing and short stubby vertical tail to make the first British swept-wing jet, soon to be unofficially known as the 'Swallow'.

The DH 108 mounted a de Havilland Goblin 3 jet engine, and a more powerful Goblin 4 of 3,738lbf (16.67kN) thrust was planned. That would have had the potential to push the DH 108 into supersonic speeds. But it lacked the aeronautical sophistication of the Miles M.52.

The first few flights were encouraging. Geoffrey de Havilland Junior, the de Havilland chief test pilot and son of the owner, gave a display flight in the DH 108 during an air show at Radlett. Then

---

9 Whittier, John Greenleaf, *Maud Muller*

10 Delta is the fourth letter of the Greek alphabet, and the uppercase Δ looks like the wing shape. A river delta is so named because its shape approximates to the triangular shape.

he took the second prototype, fitted with a longer nose and smaller canopy, on handling test flights. In a dive over the Thames estuary from 10,000ft at Mach 0.9, the Swallow went into extreme pitch oscillation and broke up. De Havilland was killed in the accident.

Now two of de Havilland's sons had been killed testing his aeroplanes. The boys' mother, Louise, found this all too much to bear and suffered a nervous breakdown. She died in 1949.

Work continued on the DH 108. Powered elevators were added as a means to control the pitch oscillations that had caused Geoffrey's death. One wonders if access to the Miles M.50 research on powered controls might have prevented the fatal crash.

A third prototype was readied for an attempt on the world speed record, which it took on 12 April 1948. It must have been a bittersweet success for Geoffrey de Havilland Senior.

Then further tests were made on controllability at trans-sonic speeds. The speed of sound was finally thought to have been exceeded by test pilot John Derry on 6 September 1948 in a shallow dive from 40,000ft to 30,000ft.

That third prototype crashed in February 1950, killing its pilot, Squadron Leader Stuart Muller-Rowland, although this time oxygen supply problems were blamed. Finally, the original first prototype of the three Swallows crashed in May of that year, killing the pilot, Squadron Leader George E. C. Genders. It had gone into an inverted spin at low altitude and although he had bailed out, his parachute had insufficient time to open.

The test pilot, Captain Eric 'Winkle' Brown, who escaped a crash in 1949, described the DH 108 as 'a killer'. During those early days of experimental jets, on average one test pilot was dying every month. Our air safety today is partly due to their sacrifice. Brown was one of the most remarkable test pilots and one of the few to survive.

He knew what he was talking about, as he had narrowly escaped a crash himself in the third DH 108. In 1949, he tested the aircraft after the controls had been fitted with power servos following de Havilland's fatal crash. Brown started a series of dives from a height of 35,000ft, then 45,000ft, when he reached Mach 0.985. But when he attempted a dive from the same height that had killed Geoffrey de Havilland, when the Mach number rose to 0.88, he experienced the same high-g pitch oscillation: 'The ride was smooth, then suddenly it all went to pieces… as the plane porpoised wildly my chin hit my chest, jerked hard back, slammed forward again, repeated it over and over, flogged by the awful whipping of the plane…'. Thinking of a way out of his predicament, Brown pulled back gently on both stick and throttle, and the oscillation 'ceased as quickly as it had started'.[11]

Brown believed that he survived the test flight partly because he was a shorter man; de Havilland had perhaps sustained a broken neck due to the violent oscillation.

Test instrumentation on Brown's flight recorded oscillation accelerations of +4 and -3gs at 3Hz. Brown described the DH 108 in his notes as, 'A killer. Nasty stall. Vicious undamped longitudinal oscillation at speed in bumps.'

Not only could Britain have broken the sound barrier, but Eric 'Winkle' Brown could well have become the first man in space and Britain could have been the nation to put him there. As we have seen, in 1945 Allied forces had raced to loot the German technology, and there was particular interest in the V2 space rocket programme. Werner von Braun, who was in command of the Vengeance weapon project, surrendered to American troops in May 1945 and was swiftly extradited to the United States. The British assembled four V2 rockets for test flights in what was known as Operation *Backfire*. They realised that the V2s had everything needed for successful

---

11 Brown, Eric, *Wings on my Sleeve*, Weidenfeld & Nicolson, London (2008)

manned space flights: a big engine, liquid pumps big enough to get sufficient fuel into it and, crucially, a sophisticated guidance system.

The designer Ralph Smith put forward a detailed proposal named Megaroc. It proposed to adapt the V2 missile into a 'man-carrying rocket'. His design involved stretching the V2s fuselage, enlarging the fuel tanks and building a pilot-carrying cockpit with two windows in place of the 1-tonne warhead. The pilot was to be Eric 'Winkle' Brown.

Space is considered to begin at 100,000m, and the manned V2 was predicted to be able to reach at least to 300,000m, although it would not have been able to get the pilot into orbit. Instead, he would have been launched into a parabolic trajectory similar to that of an artillery shell.

When slowly falling back to Earth, Brown would have experienced weightlessness and would have been the first human to see their planet from space. The cockpit capsule would have then detached from the rocket and dropped away, its heatshield protecting the new spaceman from harm. Parachutes would deploy that would float him back down to Earth. Megaroc even had a military purpose as it would have been ideal for spying on enemy territory.

'The design was totally practical,' said space historian and editor of *Spaceflight* magazine David Baker, who studied Smith's Megaroc designs. 'All the technology existed, and it could have been achieved within three to five years... By 1951 Britain could have been routinely putting people into space on a ballistic trajectory.'[12]

However, Britain was, at that point nearly, bankrupted by the war, and Smith's design was rejected by the Ministry of Supply.

In the US, von Braun was given everything he needed, and he developed the V2 into the Mercury-Redstone rocket, which took Alan Shepard into a parabolic trajectory to space on 5 May 1961.

---

12 https://www.bbc.com/future/article/20150824-how-a-nazi-rocket-could-have-put-a-briton-in-space

'Britain had been 10 years ahead of the Americans,' said Baker. 'Megaroc was essentially the Mercury-Redstone.'

Baker continued: 'It's intensely frustrating, but it's wonderful to show that this country may have been on its uppers socially, financially and with the standard of living, but we were right there with the technology.'

But the British and the Americans had both been comprehensively beaten by the Soviets, who had put Sputnik, the world's first artificial satellite into orbit on 4 October 1957, and then on 12 April 1961, three weeks before the Americans, had put Yuri Gagarin into a proper orbital flight.

Once again, the Soviets had beaten their former allies to high-speed flight. It was like the MiG-15 all over again.

Chapter 8

Not Much of an Engineer

The Rolls-Royce axial-flow Avon fitted to the later Comets was a second-generation turbojet, and eventually, after a long and painful birth, it became a successful venture for Rolls-Royce, with over 11,000 built. The Avon first ran in 1946, just as the Miles M.52 bit the dust, it went into production in 1950 and was sold as an aero engine until 1974. Industrial gas-turbine versions of the Avon for electrical power generation are still being made today, an astonishing 75 years later.

The Avon's chief designer was Stanley Hooker, whom we last met calculating the horsepower of a Whittle jet with his slide ruler. As a brilliant young mathematician from Oxford, he had been interviewed in 1937 for a job at Rolls-Royce by Ernest Hives, the managing director. After some quizzing about engines, Hives remarked, 'You're not much of an engineer, are you?' Although he got the job, the remark rankled and became the title of Hooker's autobiography.

Hives himself had come from humble beginnings and had little technical education. As works manager, he had pushed through the production of the Merlin engine during the war. He eventually became chairman of Rolls-Royce Ltd and, ultimately, Lord Hives. Vice Chief of Air Staff Sir Wilfrid Freeman said of him, 'That man Hives is the best man I have ever come across for many a year. God knows where the RAF would have been without him. He cares for nothing except the defeat of Germany, and he does all his work to that end, living a life of unending labour.'[1]

In fact, Stanley Hooker's true genius soon showed in his new job when he examined the Merlin's supercharger and realised large improvements in aerodynamic efficiency could be found. As a result, his recommendations were put onto the production line for the Merlin 45, increasing power by approximately 30 per cent. This gave RAF pilots during the Battle of Britain the vital edge they needed while they were fighting for their lives.

As we have seen, Whittle's compressor was similar to the Merlin supercharger, and so, when Hooker started working on turbojet engines, he could bring his formidable mind to the flow of gases within the new engine. But he knew at once that the centrifugal-type compressor had to go.

Hooker designed the Avon engine along the principals of the idea by Frank Whittle's nemesis, Alan Griffith: the concept of an axial compressor instead of Whittle's centrifugal compressor. He added a combustion system and a single-stage turbine that had been proved to work in the Rolls-Royce Nene engine. The rejection of the centrifugal compressor was a sign that Whittle's and Ohain's designs were truly an evolutionary dead-end; the concept of the jet engine having been proved, now the rest of the world could press on with perfecting it. However, the jet engine was about to get more complicated, just as the piston engine had before it.

The Avon started off as a single-shaft engine with a 12-stage axial compressor and a 7:1 pressure ratio (compared with 4:1 of the

---

1 Vice Chief of Air Staff Sir Wilfrid Freeman, in a letter to his wife.

Derwent V and the Nene). Development proved to be difficult, just as it had been for another great Rolls-Royce engine, the Merlin. The Avon wouldn't start, refused to accelerate and broke its first-stage blades. As Philip Ruffles wrote in the official history, 'The problems on the Avon were eventually overcome after 10 years of compressor development but this was a painful period for the Company which remained in the Corporate memory for many years.'[2]

At low speed, it was found that the flow of air through the front stages of the compressor was being throttled by the rear stages. As a result, the aerofoils were stalling, causing a flow reversal, or surge.

Hooker's colleague was Geoffrey Wilde,[3] who took over responsibility for the compressor design. He found that the problem could be solved by mounting the stators on spindles, or pivots, so they can be turned to the optimum angle. This was expensive and complex, so another approach was to bleed air off partway along the compressor, therefore increasing the velocity through the front stages. Although there was a performance penalty, this was the route taken.

In the end, it needed a 15-stage compressor and a double-stage turbine. Hooker felt he was being blamed for the problems by Ernest Hives. This was costing money.

The Avon jet was a private venture by Rolls-Royce, intended to provide engines for the English Electric Canberra. This was a twin jet-engined medium bomber designed as a replacement for the de Havilland Mosquito – a big act to follow. The name was chosen to appeal to the Australian export market, acknowledging the Australian capital city, but the Canberra proved to be so popular that it was exported to 14 other countries. No less than 1,352 Canberra aircraft were built in several factories around the world. Even the Americans bought it and built it under licence as the Martin B-57 Canberra.

---

2 Ruffles, Philip, *The History of the Rolls-Royce RB211 Turbofan Engine*

3 Wilde went on to design the Rolls-Royce RB211 jet engine.

Over 400 of these were built and it held the distinction of being the first jet bomber in US service to drop bombs during combat.

Eventually the Avon was made to work, and once again Rolls-Royce named a turbojet engine after a river. But river Avon is a tautology: Avon means 'river' in the British Celtic language, and so translates as 'River River'.

At first, the Rolls-Royce Avon engines provided 6,700lbf (30kN) of thrust, but later versions of the Avon provided an impressive 12,690lbf (56.4kN), and 16,360lbf (72.8kN) with reheat, or afterburner. The all-important pressure ratio was 7.45:1.

By the end of the war, Rolls-Royce had decided that its existing piston engines were an evolutionary dead-end and moved all future jet work from the old Rover Barnoldswick factory to Derby, its main engine site. Hooker felt his role in the company was diminished – he had never been appointed Director of Engine Development and, after an angry falling-out with Hives, he resigned. It was to be a great loss to Rolls-Royce but a great acquisition for the Bristol Aero Engine company, whose development team he then joined.

When reading British engineering history, it's hard to avoid narratives of inventive genius endlessly frustrated by unsympathetic management: first Whittle, now Hooker. Was there any truth in it?

The managers generally had a wider view of what was needed, whereas the engineers concentrated on the problems at the workbench. It's hard to say, but history is written by writers, and writers generally favour the solitary genius.

The Rolls-Royce Avon engines were mounted halfway along the Canberra's wings. Each engine drove a 6kW generator for the aircraft's 28-volt electrical system, a hydraulic pump and a bleed air system for cabin pressurisation, just like the Comet jet airliner. Coffman 'shotgun' engine starters were employed. These extraordinary devices consisted of an explosive cartridge of cordite, detonating in a cylinder containing a piston mounted on a coarse screw-thread that spun the engine. They had also been used for starting Rolls-Royce Merlins and many other engines.

The Canberra bomber was a straight-wing design and, as a result, it had excellent high-altitude abilities: for most of the 1950s, the Canberra could fly at a higher altitude than any other aircraft in the world, evading any jet fighter sent up after it. The early prototype operated by Rolls-Royce during Avon tests would regularly fly to 63,000ft, where the useable speed range was only 25 knots, and, in 1957, a Canberra established a world altitude record of 70,310ft.[4] The only aircraft that could beat the Canberra for altitude was the later Lockheed U-2 spy plane. Both aircraft flew in a small performance envelope named 'coffin corner'.

I knew a Lockheed U-2 spy plane test pilot who told me about life in coffin corner. Jerry Hoyt just loved aluminium, or 'aluminum'[5] as he called it. When I first met him, he was lying under the aluminium hull of his sailboat in Florida, polishing it hard. Later he gave me drinks in cans made of the same metal, sitting in the sun outside his aluminium Airstream trailer. He explained how the CIA had used the U-2 high-altitude reconnaissance aircraft for flights over hostile countries in the late 1950s. It was made, of course, of aluminium, and ran on cigarette lighter fuel: LF-1. 'Nothing else would burn at

---

4 In February 1951, a Canberra set another world record when it became the first jet aircraft to make a non-stop transatlantic flight. The aircraft set several records for height, speed and distance during the 1950s.

5 In 1808, Sir Humphry Davy identified the existence of the metal in alum, which he at first named 'alumium' and later 'aluminum'. So it was an Englishman who christened the metal alum*inum*, the American way. The official name, 'alum*inium*', was adopted to conform with the -ium names of most other elements.

that height; he said. 'That bird was a giant flying Zippo.' Jerry Hoyt told me how he set several time-to-climb world records in his U-2 on 17 April 1989.

He took off just after dawn with an apocalyptic roar down the runway. The U-2 was airborne in only 150ft and Hoyt hauled back on the stick and put the aircraft into a near-vertical climb. The General Electric F118-101 turbofan engine, with 17,000lbf (76kN) of thrust, shot Jerry skywards like a rocket. As the U-2 passed through each 10,000ft mark, it broke yet another climb-to-altitude record, cutting each of them by half: just one of them was the time to climb to a height of 65,616.8ft (20,000m) in 12 minutes 14 seconds. He reached 73,736.9ft in level flight 14 miles (23km) above the Edwards runways on that glorious morning in 1989.

To stay at 70,000ft during reconnaissance spy flights, Jerry's U-2 had to fly close to its never-exceed speed of 540mph (870km/h).[6] The only problem was that at that height, the U-2 would stall at 530mph (850km/h). This narrow margin of 10 knots is called 'coffin corner' on the graph because breaking either limit would cause airflow separation at the wings and initiate a stall.

If Jerry stalled, the U2 would tumble out of the sky, overstress the airframe and break up. For most of the time, on a typical mission, his U-2 was flying less than 6mph (10km/h)[7] above stall speed. 'I had a realistic margin of error of only about 2 knots,' he told me.

Jerry knew Gary Powers, the U-2 pilot who was shot down by a Soviet surface-to-air missile in May 1960. Pilots carried a suicide pill on their missions. This 'L-pill' contained liquid potassium cyanide and, if crushed between the teeth, would ensure death within ten seconds. After one U-2 pilot nearly swallowed the cyanide pill by mistake, thinking it was a sweet, it was stored in a box. But then the CIA realised that if a pill was accidentally

---

6 Mach 0.715 (470 knots, 540mph, 880km/h) at 72,000ft.

7 5 knots (6mph, 9km/h).

broken inside the cockpit, it would kill the pilot by inhalation, so the L-pills were removed. As a replacement, the CIA Technical Services Division developed a needle poisoned with saxitoxin, a powerful shellfish toxin, hidden in a silver dollar. Jerry probably would have preferred aluminum.

The Canberra was built by English Electric, a company formed after the armistice of the First World War from an amalgamation of five businesses that, during the war, had been making munitions, armaments and aeroplanes. During the Second World War, they had made the Napier Sabre, probably the best piston aero engine of the war, plus 3,000 aircraft, tanks, locomotives, submarines and ships. Because de Havilland's was busy making Mosquitos, English Electric Aircraft also undertook production of the Goblin jet-engine powered de Havilland Vampire at its factories in Preston, Lancashire. The Vampire was the RAF's second jet fighter after the Gloster Meteor, and the first to be powered by a single jet engine. So English Electric was well placed to make successful jet aircraft. The public was well aware of the name, with English Electric washing machines proudly displayed in many 1950s homes.

Rolls-Royce's new Avon and the sheer bravery of the test pilots were enough to keep Britain ahead in the jet-engine technology race. The big advantage over the previous Nene engine was Hooker's adoption of the axial compressor, which not only allowed for a smaller engine diameter, allowing thinner fuselages, but also provided greater thrust: this single engine gave the same power as the two Rolls-Royce Derwents of the Gloster Meteors.

One aircraft manufacturer that immediately saw the advantages of the new engine was Hawker's, famous for its wartime Hurricane fighter. In March 1948, the Air Ministry issued a Specification F.3/48, which called for a trans-sonic interceptor fighter with a speed of 629mph (1,010km/h) at 45,000ft and a high rate of climb, while carrying an armament of four 20mm (0.79in) or two 30mm (1.18in) cannon. It was clearly thinking of the Soviet Union's nuclear ambitions, which were realised by its first successful atom bomb test the following year. Soviet bombs would be delivered by high-altitude jet bombers, and, once again, Britain's defence would be its fighters.

Hawker's designer, Sydney Camm, came up with an attractive swept-wing design with neat air intakes either side of the fuselage. The slender Avon engine kept the fuselage slim. The Hawker Hunter took the world airspeed record on 7 September 1953, with a speed of 727.63mph (1,171.01km/h) over Littlehampton, West Sussex. This world record stood for less than three weeks before being broken by 10mph (16km/h) on 25 September 1953 by the Hunter's rival, the Supermarine Swift. This aircraft also used the Rolls-Royce Avon. The pace of jet technology was hotting up.

To showcase these exciting jet planes the British public was invited to see them in action over their heads at the biennial Farnborough Air Show. And what they wanted to hear was the sound barrier being broken.

De Havilland used two of the new Avons in its DH.110 Sea Vixen. This was a twin boom-tailed two-seat carrier-based fleet air defence fighter built for the Royal Navy. The DH 110 had the distinction of being the first British two-seat combat plane to achieve supersonic speed, and at the 1952 Farnborough Air Show it demonstrated just that. The test pilot, John Derry, took a low pass just over the runway and broke the sound barrier, following this with a steep climb.

Then the aircraft disintegrated.

The leading edge of a wing fractured, the swept wings ripped off, the fuselage fell apart, and the two Avons, still spinning, were filmed flying through the air like smoking projectiles. One of them landed in the crowd. Cockpit debris hit the crowd, too. Thirty people died that day, including John Derry and his flight-test observer, Tony Richards.

No one sued.

Astonishingly, the airshow continued. Test pilot Neville Duke took the controls of a new Hawker Hunter, took off, dived – and broke the sound barrier.

Chapter 9

707

The tyranny of distance had dictated the choice of aero engines for American aircraft during the Second World War. The breadth of the continent and the vastness of the Pacific Ocean demanded fuel-efficient piston engines with a long range. The US Navy preferred large-capacity air-cooled engines because there was less to go wrong, and so that's what they got.

Timber dealer Bill Boeing[1] had gone to the Pacific Northwest in the early 1900s to check out the timber prospects, which turned out to be pretty good. Back then, Seattle was a rough port town. Bill Boeing became interested in the new craze of aviation and began building seaplanes out of spruce. This wood is particularly good for airframes because it is light, stiff and has great resistance to splitting. Due to the favourable strength-to-weight ratio it was – and is – used extensively in aircraft construction for spars and longerons.

So 'Built Where the Spruce Grows' became the Boeing Airplane Company motto.[2] Later, when Jerry Hoyt's aluminum replaced spruce, the Pacific Northwest was still a good place for aircraft materials: the cheap hydro-electric power produced large amounts of the metal.

In 1933, the Boeing aircraft company had built an innovative modern ten-passenger all-aluminium airliner with retractable undercarriage, de-icing equipment, autopilot and twin Pratt &

---

1 The name was an Anglicised version of the German name Böing.

2 Its later replacement, 'Boeing build something better!' only makes you wonder, 'If they do, then where is it?' The earlier 'If it's not Boeing, I'm not going' was not much better. And Airbus's 'We make it fly!' invites the rejoinder, 'Well, I certainly hope you do.'

Whitney Wasp 9-cylinder air-cooled radial piston engines. The Boeing 247 was not the first all-metal airliner – that accolade goes, of course, to the Germans – but it was regarded as a huge advance for the American market.

Then the Boeing's limelight was stolen by the 14-seat Douglas DC-2. Donald Douglas copied the Boeing and made it 25mph (40km/h) faster, which meant an hour saved on the New York-to-Chicago run.

The DC-2s legendary successor, the 21-seat DC-3, was even more successful, with 16,000 examples built, some of which are still flying today. The DC-3 was fast and comfortable and suddenly made air travel popular across the United States. The 9-cylinder Wright R-1820 Cyclone 9 air-cooled engines were economical and gave a long range. As a result, transcontinental flights could now cross the US in 15 hours with only three refuelling stops whereas, previously, passengers had to make short hops by air during the day, followed by railway sections of the journey by night. A sleeper version, the Douglas Sleeper Transport (DST), even enabled 16 horizontal passengers to snore their way across the fly-over States. The Douglas DC-3 out-sold the Boeing 247 by 200:1 and in 1939 carried 90 per cent of the world's airline passengers. During the war, it was converted for military transport purposes and named the Douglas C-47 Skytrain or Dakota.

Boeing had made headlines just before the war with its double-decker B-314 Clipper transatlantic flying boat, and this is how: the far-sighted Juan Trippe, the founder and chairman of Pan American World Airways, had always dreamed of a commercial service between the New World and the Old. British Imperial Airways and Pan American Airways had experimented with transatlantic flights in 1937, the British using the Short Empire S.23 flying boats, which took a day to make the crossing.

Trippe realised the main stumbling block was not technical but political, and so he negotiated for landing rights in Newfoundland, Greenland, Iceland and Northern Europe. Meanwhile, he requested bids from eight aircraft manufacturers for a 100-seater airliner. Boeing won the competition with its B-314 flying boat. It was enormous, the biggest commercial airliner in the air until the arrival of the Boeing 747 'Jumbo Jet' 30 years later. It had a range of 3,500 miles (5,633km) and could cross the Atlantic or Pacific oceans.

Pan Am started carrying passengers between New York and Southampton on 8 July, just in time for the beginning of the war. The service was for the super-rich, with passengers paying US$760 for the round trip (around US$14,200 today). For this they received the most luxurious service in airline history, with separate dressing rooms, fold-down beds and six-course meals cooked by restaurant chefs and served by white-coated stewards in a separate dining room. Even so, each passenger was weighed before stepping aboard, a good idea that was eventually rendered unnecessary by the sheer power of jet engines.

In 1943, the Lockheed[3] Aircraft Corporation revealed the Lockheed 'Connie' Constellation, a luxury four-engined airliner with 18 sleeper cabins and a distinctive triple tail like Shiva's trident, which was low enough to sneak into existing hangers. It also had the most beautiful dolphin-like silhouette. Lockheed had worked on the design with Howard Hughes, the owner of TWA, and who was another one of those visionary and extraordinary personalities in aviation.

When we think of Hughes today, we might think of the reclusive billionaire, suffering chronic pain due to the injuries sustained by several air crashes, and tortured by his obsessive-compulsive disorder. He ended his life sitting naked in his private cinema, watching endless re-runs of the film *Ice Station Zebra*, letting his hair and fingernails grow unchecked and eating the same dish, day after day.

---

3 Allan Loughhead had changed the spelling of his Scottish surname to Lockheed in 1934.

But Howard Hughes was also obsessed with aircraft and had set many speed and distance records. Made independently wealthy by his father's invention of the oil well drill bit, he had, in turn, been a professional golfer, real-estate investor and film director – he first came into the public eye as Jane Russell's boyfriend. While directing *The Outlaw* he became obsessed by the starlet's breasts and, using his aero-engineering knowledge, designed a special cantilevered brassiere to reveal more of them. Russell found the contraption ridiculous and uncomfortable and secretly discarded it before filming, wearing her own bra.

Hughes owned the RKO Hollywood film studios and then got into the aviation business. He survived no less than four self-piloted airplane accidents: one nearly fatal. This last one happened when he was testing his twin-engined Hughes XF-11 reconnaissance aircraft from the Hughes airfield at Culver City, California. An oil leak caused one of the starboard contra-rotating propellers to suddenly slam into reverse pitch. As with most twin-engined aircraft, this was a terminal problem. The XF-11 yawed and lost height, crashing into the Beverley Hills district and demolishing a house. Even Hughes' air crashes were high rent. He was nearly killed.

Hughes invested heavily in the Lockheed 'Connie' Constellation. It offered a luxury service with fresh flowers, sleeper cabins, beds with linen sheets, and elaborate meals. The piston-engined Connie was capable of flying across the Atlantic, and the TWA's first transatlantic flight departed from Washington, DC, on 3 December 1945, and arrived in Paris on 4 December after refuelling stops at Gander and Shannon.

But it wasn't cheap. The tickets cost US$300 to $400 ($4,275 to $5,700 at today's values). It should be remembered that jet travel was still the prerogative of the rich. It wasn't until 1952 that the first economy class was launched, by TWA. And, ironically, it was to be the low-cost carriers that would eventually see the demise of TWA.

Howard Hughes had also helped Boeing to develop its 1938 Model 307 Stratoliner, which was the first airliner to offer a pressurised cabin, allowing it to cruise at an altitude of 20,000ft, above the bad weather. But Boeing had been comprehensively beaten by Douglas and Lockheed in the pre-war commercial aviation market.

During the Second World War, though, the company roared back with its successful B-17 Flying Fortress and B-29 Superfortress bombers, making huge wartime profits. It was these profits, together with Boeing's dismal post-war commercial sales, that lead to a tax loophole that, cleverly exploited, made Boeing the world's largest maker of jet airliners today.

This is how they did it:

Just after the Second World War, Boeing had lost millions on its Clipper and Stratocruiser airliners, again beaten by Douglas and Lockheed. Like the Clipper, the Stratocruiser majored in luxury: it had Pullman sleeping berths, separate women's and men's dressing rooms and a huge cocktail lounge. Maxim's of Paris provided the seven-course meals, which began with champagne and caviar and were served on china. Four Pratt & Whitney R-4360 B6 Wasp Major engines, with 28 cylinders each, gave a speed of 350mph (560km/h) and a huge payload of 30,000lb. However, the engines and propellers were excessively complex; there were several engine overspeeding incidents and the aircraft suffered 13 hull-loss crashes, with 139 dead. In the end, the Stratocruiser (which unfortunately looked like a bulldog) just could not compete.

As a result, by 1950, Boeing only had 1 per cent of the commercial market. But then, during the Korean War, Boeing made huge profits on the fleets of bombers it sent over the Pacific. It was boom, bust and then boom.

During the Korean War, the US government suddenly slapped an excess profits tax on armaments companies. Excess profits would consist of those made over and above what the company had made during the peacetime lull of 1946 to 1949. For Boeing, who had hardly made any profit during that time, and then lots of profit during the Korean War, this tax was a disaster. It stood to lose 82 cents in every dollar of profit made after the Korean War, whereas Douglas Aircraft was to be billed 64 cents and Lockheed only 48.

Then, Bill Allen, the Boeing president and former tax lawyer, had a brainwave. If he could spend a large amount of money on developing a new aircraft, it would count as a legitimate business expense.

He had been watching the rise of the de Havilland Comet; he had visited the British factory and thought Boeing could do better. This was a bold move as Boeing had no customers for a jet airliner and the technology was new, but for every dollar Boeing spent, only 18 cents would have been theirs to keep anyway. Douglas Aircraft and Lockheed weren't in such a favourable position.

This was a brilliant move, and it would make Boeing the world leader in jet aircraft for decades. Allen then had another idea, which was to develop the new jet airliner in parallel with a four-engined jet tanker, which was needed by the USAF to refuel the new Cold War jet bombers. The two types of aircraft could be developed from the same prototype, which would be called the Boeing 367-80, or 'Dash-80'. This would demonstrate the advantages of jet propulsion to a sceptical American market.

And the Americans *were* sceptical about jets. After the Korean War experience, they were regarded as too fast, too thirsty, smoky and loud – thunderous engines of war. They would be too fast for air traffic controllers to track on their radar screens: 'There is a limit on how many blips a single brain can safely juggle,' fretted the *New York Times*. No one could imagine them being domesticated enough to operate at airports. And jets were indeed thirstier than the piston-engined competition. No one by then had proved that a jet airliner could make it across the Atlantic non-stop, and if a jet had to stop for fuel, it lost its main competitive advantage: speed.

Also, the American airlines had invested millions of dollars in the new post-war generation of piston-engined aeroplanes such as the Douglas DC-3 and Lockheed Constellation. They were making huge profits flying these aircraft in the buoyant economy and so were reluctant to take the plunge with new jet technology.

There was a parallel in the automobile market. Although American cars were growing bigger in the 1950s, with flashy chrome and tailfins, underneath they still had the old cast-iron engines and the cart springs of their predecessors. When the British and the Italians started producing alloy V12 engines and independent suspension, the US makers never thought they could afford to develop them. Cheap, cheerful and reliable was the American way.

The cost considerations made American airlines reluctant to invest in a jet airliner: 'The only thing wrong with the jet planes of today,' said TWA's boss in 1950, 'is that they won't make any money.'

As a result, none of the American aircraft makers were willing to make the huge investment necessary to build a jet airliner alone, although Boeing had clearly been thinking about it for a while. In 1949, its employees' magazine had complained:

> Government financial aid will be required if we are to overtake and pass the subsidized British aircraft industry in its bid for domination of the future jet transport field... No company can risk its capital in the building of a wholly new type of airplane, such as the jet transport.

Then, Allen bet the company and made his move. And in 1952, just weeks after the Comet made its inaugural flight, the Boeing board signed off US$16 million ($155 million today) to build a jet-engined prototype that would be called the Boeing 367-80, or 'Dash-80'. It would lead to the development of the Boeing 707 passenger jet airliner for the domestic airlines and the KC-135 Stratotanker aerial refuelling tanker aircraft for the military.

The 707 jet airliner promised a great deal: it would fly at 600mph (970km/h) instead of the old piston-engined Stratocruiser's 350mph (560km/h). It would carry up to 120 passengers instead of just 80. It would fly as high as 50,000ft instead of 30,000ft, and it would reach

Europe in six hours instead of 11. Like any good car dealer, Bill Allen didn't mention the enormous fuel consumption and the deafening noise of turbojets. In building the new jet without an airline partner to shoulder some of the cost, Boeing was taking a big gamble.

The jet engines for the prototype would be Pratt & Whitney J57 axial-flow turbojets based on the Griffith design concept. As the pressure ratio helps efficiency and thus fuel consumption, this engine had no less than nine stages in a first low-pressure compressor section. The blades decreased in size and spacing as the air was compressed further and further. These are quite beautiful to behold, and retired compressor wheels often reappear as coffee tables in high-rent penthouses.

Then in a high-pressure section, seven stages compressed the incoming air furthermore to 12.5 times its original pressure. Whittle could only manage a pressure ratio of 4:1 with his single impeller.

The combustion chambers, or combustors, on the J57 were also tuned for efficiency; as we have seen, there are huge demands on these components in a jet engine. They must accept high-pressure air at high velocity and burn the fuel injected into this whirlwind without it being blown out. If there *is* a flame-out, the combustors must be able to re-light the engine at high altitude. The combustion itself must be complete, with no ugly inefficient smoke. And the combustors must withstand high temperatures yet be small and light.

Whereas the Whittle engine had separate 'cans' containing the fuel nozzle, igniter and air passages, the combustor on the J57 were 'cannular', a portmanteau word. These had combustion areas in can liners with nozzle and igniter, but they were mounted in a single ring-shaped annular casing sharing the same high-pressure air from the compressor stages. These were lighter than cans and suffered from less pressure drop. However, they were difficult to maintain.

Finally, the J57 had a single-stage high-pressure turbine followed by a two-stage low-pressure turbine. The whole design effort was directed at extracting the utmost from every drop of kerosene. The final performance figures were impressive: maximum thrust was 12,030lbf (53.5kN), and the all-important specific fuel consumption was 0.785lb/(lbf·h).

The Boeing 707 had what is called second mover advantage. Although the British had beaten them to the title of 'first jet airliner', Boeing could learn from the Comet's successes and its mistakes. The British had also done the heavy lifting in persuading the flying public that jets were a good idea. But, like the first penguin that dives off the ice floe in front of the rest of the flock, the British had paid for their audacity.

The 707's cabin would be like a spaceship, with subdued lighting, individual controls for ventilation, modernist-style seating and just a distant whine from the jet engines out on the wings. Only the public living around the aerodromes would hear the full roar of the turbojets.

For the airframe of the Boeing 367-80, or Dash-80 prototype, Boeing experimented with swept-wing angles in its huge wind tunnel at Seattle, eventually settling on 35°, whereas the Comet had 20° of sweep. The designers placed the four Pratt & Whitney J57 engines in pods hanging down from the wings on stilts instead of being buried in the wing roots as on the Comet. There was clever design at work here: the twisting moment of the podded engines around the wing allowed it to resist aerodynamic loads, meaning that the wing structure could be lighter.

This plane was not going to be as elegant an aeroplane as the British jetliner, but the wings would be easier to build. The windows were rounded, and instead of the beautiful but expensive dolphin shape of the Constellation, the future 707's fuselage was going to be a simple tube. This was an aircraft that could be built by the mile and sold by the foot.[4]

The new aircraft was big, too: the 707 airliner decisively outranked the British jet and the Lockheed Constellation. The Connie could only carry 66 passengers and took 14 hours to cross the Atlantic. The new Boeing would be able to halve the time to seven hours, travelling at 600mph (970km/h), and carry twice the number of passengers at 120. It would carry three times as many passengers as the Comet.

This raised an important question: if there was a major accident, it would immediately become the biggest disaster in aviation history with up to 140 dead. Could Boeing survive such an event?

'It was a question of policy as to whether you should expose one hundred people to the hazards of commercial aviation, and whether Boeing could survive the headlines of a hundred people being killed in a single crash,' remembered Maynard L. Pennell[5], a senior executive of the commercial development division.

That same question would come to haunt Boeing a couple of years later. In our own times, it has come to trouble the company again, when two of their 737 Max airliners crashed in 2018 and 2019, killing 346 people and leading to the grounding of the whole fleet of 790 aircraft for over a year while their controversial MCAS flight-control system was reviewed.

---

4 Parked nose to tail, all the Boeing 707s ever built would stretch for over 26 miles (42km). And the Boeing 737 would stretch over 240 miles (385km).

5 Pennell went on to manage Boeing's SST supersonic transport programme, engaging in a government-sponsored contest against Lockheed for the right to manufacture the airframe of the faster-than-sound plane. By mid-1966, he had unveiled a model of a 300-passenger, 330ft-long aircraft (100m) meant to fly at 1,800 miles (2,900km) an hour, almost three times the speed of sound, with a range of about 4,000 miles. Boeing than had to watch as the supersonic Concorde, a joint venture between Britain and France, took to the air.

In fact, jet airliner travel was to become safer and safer thanks in part to the reliability of jet engines and their lower-flammability fuel.

At first, no American airline was willing to risk millions of dollars on the new Boeing design. Then the de Havilland Comet began flights in 1950. Much to the chagrin of the American aircraft companies, the British had beaten them to the draw, and they were reaping the rewards. On some Comet routes, there was a month's waiting list for tickets.

An article in *Harper's* magazine[6] envisaged a Comet standing at Gate Two at New York's Idlewild Airport in 1953, and a piston-engined American airliner standing next to it at Gate Three:

> If you are by then the president of an American airline and happen to be standing at Gate Two or Three, the chances are you will be boiling mad, for the British plane will be packed and yours will be almost empty. If you are an American aircraft manufacturer you will recall nostalgically the good old days of 1949, when piston-engine American transports were the envy and the standard of the world's airlines – the fastest, the safest, and the most efficient to operate. Fantastic? Probably not. For jet transportation is here today. The British do have an acceptable jet transport… These were of the clan once considered here in America to be five years behind the times. Today they puff their pipes and cheerily evade questions about advanced British design.

At first, Boeing could not find a buyer for its new airliner. To add a twist to the knife, a second jet passenger plane prototype[7] took to the skies two weeks after the Comet. This was the Avro Canada

---

6 March 1950.

7 In the event, the Avro C102 was cancelled due to a lack of orders.

C102 Jetliner. The company had been set up during the war as a shadow factory that built Lancaster bombers for the British A. V. Roe company. Now it had built a prototype four-engined jet with 50 seats. The engines were Rolls-Royce Derwent V centrifugal-flow turbojets, each of 3,600lbf (16kN) thrust.

Worse still, the Soviets were rumoured to be working on a twin-engined jet airliner, which would appear in 1955 as the Tupolev Tu-104.

Something clearly had to be done. Bill Allen of Boeing decided to demonstrate his new prototype in August 1955 at the Gold Cup races on Lake Washington, Seattle. The arrival would be unannounced in front of hundreds of aviation industry executives.

During a lull in proceedings, the crowd's attention was drawn to the sight of a huge four-engined jet roaring into the air. The single Dash-80 prototype was being flown by the brash and flamboyant 'Tex' Johnson, the Boeing chief test pilot. It gained height and approached Lake Washington, with everyone's eyes fastened upon it. Then, as it thundered over the lake, one wing suddenly dipped and the whole jet airliner slowly started to turn over.

Down on the lake, Bill Allen's stomach turned over with it. Up there were all his company's future hopes, plus US$18 million of investment: the only prototype. And it looked as though it was just about to crash into the crowd, killing thousands. Just like the Farnborough Airshow.

The brightly painted airliner turned completely upside down, then continued to complete the barrel roll. The crowd gasped with amazement. Then the aircraft turned around, came back to the lake and did it all over again.

Bill Allen was silently furious. This wasn't in the plan at all, but he had to pretend that he knew that it was going to happen…

Reckless or not, Tex Johnson's barrel roll certainly got the aviation industry talking. Two months later Juan Trippe's Pan Am put in an order for 20 Boeing 707's. Not only that, but he would also buy 25 Douglas DC-8s, when that suspiciously similar four-engine jet airliner was ready to buy off the stocks. Donald Douglas had done it again – he had copied a successful Boeing design. The DC-8 even used the same engines, the Pratt and Whitney J-57s.

One has to admire Trippe's courage – to buy one untried design was unusual, but to buy two was unheard of.

Other airlines clamoured to follow suit; suddenly everyone wanted jets. Competition was ferocious. In November 1955, American Airlines announced it would take 30 Boeing 707s and, by the end of 1956, the Boeing 707 was beating the Douglas DC-8 three to one. By the beginning of 1958, Douglas had sold 133 DC-8s and Boeing had sold 150 707s.

Then, in 1959, President Eisenhower's White House ordered three 707s to become the first jet Air Force Ones. Eisenhower was painfully aware that Russia's Nikita Khrushchev had beaten him to flying in a jet, the Tupolev 104. This aircraft had two Mikulin AM-3 turbojets mounted in the wing roots, rather like the de Havilland Comet. They produced 19,176lb thrust. The cabin interior was a curious confection of mahogany, copper, glass and lace. It looked rather like a Victorian brothel, but it was the only jet airliner flying from 1956 to 1958 after the Comet was grounded. It had a poor safety record due to a habit of stalling when landing, and 16 out of 96 aircraft were lost in accidents.

But, just as with Sputnik, Russia did the presidential jet first. And at last, America had joined the Jet Age.

Chapter 10

The Jet Set

O n 4 October 1958, a BOAC de Havilland Comet 4 became the first jet airliner to fly paying passenger services across the Atlantic, operating in both directions between London and New York. In achieving this feat, the UK long-haul flag carrier beat its great US rival, Pan Am, which was planning to be the first airline to operate a jet across the Atlantic with the Boeing 707.

With a shrill scream from the Rolls-Royce Avon turbojets, the revised Comet climbed into the sky from London Airport. 'It was marvellous,' recalled BOAC stewardess Peggy Thorne. 'We were used to travelling to New York on Boeing Stratocruisers which took up to 20 hours. We couldn't believe the flight was possible in such a short time. Our customers loved it – they ate and drank from when they got on board until the time they got off.'[1]

Up in the cabin was Captain Hugh Dibley. 'Pan Am perceived themselves as the world leaders,' he remembered. 'They just assumed they'd be the first across the Atlantic with the 707. BOAC Comet 4's debut really took the wind out of their sails, which upset them enormously.' Shortly after take-off, another BOAC Comet started out from New York in the reverse direction.

That city's local newspaper, the *New York Times*, had the headline 'British Gloat as Their Comet Wins Race to Inaugurate Transatlantic Jetliners' and *The Times* of London proclaimed modestly, 'The New Age of Travel Begins'.

---

1 https://www.flightglobal.com/networks/video-recollections-of-how-boacs-comet-won-the-transatlantic-jet-race/129743.article

Harold Macmillan, the British prime minister, was delighted: 'The whole nation takes pride in the fact that a British aircraft has led the world.' And privately, he sent a note to Sir Geoffrey de Havilland. 'Heartiest congratulations to you on your magnificent effort in getting the Comet 4 first off the mark in regular passenger service across the Atlantic,' he wrote. 'The resurgence of a Comet airliner is a fitting reward for faith in the future of this fine product, and the whole nation takes pride in the fact that a British aircraft has led the world into a new turbojet age.'

Geoffrey de Havilland replied with feeling: 'The Comet has meant more to me than any other aeroplane we have built. It has been the most challenging, the most disheartening and finally the most technically rewarding project we have ever tackled.'

Aviation Week said there was 'no more traditional British characteristic than dogged determination to push on toward a goal despite any and all adversity that appears along the way… there is no better example of this British determination than the reappearance of the de Havilland Comet in the livery of the British Overseas Airways Corporation on the world airlines after an absence of several years.' It was 'indeed a bold stroke that provided British aviation morale a much-needed boost.'

However, it was a Pyrrhic victory.

'With another victory like that we'll lose the war,' Pyrrhus of Epirus had said gloomily. And so it was for the British aviation industry. A long strike by bus workers shut down London Airport. When the Boeing 707 started flying, the triumphalism ebbed away. The Comet had half the Boeing's range and could carry only half the number of passengers. 'When the 707 came along there was no comparison economically with the Comet and of course the 70 could carry over twice the number of passengers,' said Captain Dibley. 'So the life of the dear old Comet was limited.'

Soon, BOAC stopped flying Comets across the Atlantic and bought Boeing 707s for the route. The Comet 4 was used for medium-distance routes in Africa, South America and Asia. Dan-Air flew Comets until 1980 on European routes, and the last Comet flight was in 1997.

However, the first Boeing 707 flight to Paris was a reminder of the limitations of technology. First the thirsty jet had to stop at Gander, Newfoundland, to refuel. Then thick fog delayed landing at Paris for over an hour and a half, making the new jet not much faster than the TWA's Super-Constellations. And the only stars on board were the actress Greer Garson and the bunion millionaire, Dr William Scholl.

And so it was going to be the Boeing 707, not the Comet, that carried the Jet Set.

Come fly with me, let's fly, let's fly away
If you can use some exotic booze
There's a bar in far Bombay
Come fly with me, let's fly, let's fly away.

Frank Sinatra's *Come Fly with Me* album of 1958 captured the zeitgeist, but his crooning lyrics disguised an inconvenient truth: Sinatra was terrified of flying. Even though he sang the anthem that defined the Jet Set, during flights he used alcohol and distractions to take his mind off his fear. 'Live every day like it's your last,' he would say cheerfully, 'then one day you'll be right.' He had a recurring nightmare about the mysterious death of big band leader Glenn Miller, whose aircraft taking him from wartime Britain to France disappeared somewhere over the Channel.[2]

Before the Jet Set was the Yacht Lot, a dissolute assortment of international aristocrats, showbiz stars, dubious nouveaux-riches and random Euro-trash hangers-on. There were wealthy socialites

---

2 Miller's aircraft was a Noorduyn Norseman, notorious for carburettor icing. The weather was cold and foggy, likely to exacerbate this. The aircraft probably ditched and drowned the occupants. Turbojets do not suffer from this particular problem.

such as Joseph Kennedy and Barbara Hutton of the US, Aristotle Onassis and Stavros Niarchos of Greece, Gianni Agnelli of Italy, Aly Khan of France and 'Baby' Pignatari of Brazil, all of whom holidayed on luxury yachts. They allowed selected individuals to travel with them, such as Frank Sinatra because he was the coolest swinger in town, and Porfirio Rubirosa because he had married the two richest heiresses of the era, Doris Duke and Barbara Hutton, the Woolworth heiress. Women were usually decorative except for these two, who were rich enough to own their own yachts.

I spent a night on Barbara Hutton's yacht to try to capture the flavour of the time. I slept in the owner's cabin, and on the wall was a photograph of her with her third husband of seven: Cary Grant (the couple were dubbed Cash and Cary). Hutton was the original 'poor little rich girl' who never seemed to find lasting happiness with any of her men. The German-built[3] yacht was an 18th birthday gift from her father and was quite simply a floating palace, with dozens of servants, the finest of wines and gourmet food. The captain's lounge had a glass floor so the rich could observe the engine-room staff slaving beneath their feet, and so for those below, their glass ceiling was an actual physical and social barrier. These were the standards that the Jet Set were going to demand in the first-class cabins of the Boeing 707s.

Federico Fellini's film *La Dolce Vita* (1960) sums up the superficial and extravagant lifestyle of the Jet Set. Marcello is a cynical gossip columnist who pursues rich people, seeking ephemeral pleasure in parties and sex. When a famous film star comes to Rome on a jet airliner, he does everything he can to meet her. But he finds he has been drawn to an image, not a reality. The film coined the name 'paparazzi', the term for celebrity-stalking photographers, named

---

3 This yacht was bought by the Admiralty in 1940 and renamed *Troubadour*. However, the German-built MAN diesel engines needed spare parts, so the vessel was laid up in Lamlash Bay, Isle of Arran. After action on the Normandy beaches and at Walcheren, my RN lieutenant father was billeted on the luxury yacht. It was a tough posting but someone had to do it.

after the hero's camera-toting sidekick, Paparazzo, whose name, in turn, was borrowed from a hotelier in George Gissing's travel book, *By the Ionian Sea. La Dolce Vita* became synonymous with the godless and decadent lifestyle of the Jet Set. The opening scene, with a statue of Christ flying over Rome, suspended beneath a helicopter is a parody of Christ's second coming, which the Vatican found offensive. The film was banned.

The Jet Set was a fascinating cocktail of celebrities, jet technology and glamourous surroundings. Among them were Frank Sinatra's Rat Pack buddies: Dean Martin, Sammy Davis Jr, Peter Lawford and Joey Bishop, jetting to destinations such as Acapulco, Nice and Rome. The newspapers were full of their orgies and indiscretions, and films such as *An American in Paris* (1951), *Roman Holiday* (1953) and *Breakfast at Tiffany's* (1961) whetted the American public's desire to join them.

Then the fictional character who did most to personify the Jet Set landed at Kingston, Jamaica. 007 had just arrived – on a 707.

The shot of the Pan American 707 making a perfect landing in *Dr. No* (1962) was a piece of product placement that could not have pleased Juan Trippe of Pan Am any more than if he had paid for it – which he hadn't.

The author of the James Bond books, Ian Fleming, was himself an upper-class spy, epicure, womaniser, alcoholic and hobby sadist. He longed to be a respected travel writer like his older brother, Peter Fleming. But that post was taken, so he turned his talents to depicting an exaggerated version of his own life. For he too ate in gourmet restaurants, stayed in luxury hotels and never went to bed sober.

Fleming was a Royal Navy commander working for naval intelligence during the Second World War, and James Bond was a Royal Naval Reserve commander in MI6. Fleming was cynical and intelligent; Bond added coldness and murderousness. Fleming's choice of alcohol was precise: 'Martini, three measures of Gordon's, one vodka, half a Kina Lillet, shaken, not stirred, served in a champagne goblet.' Bond orders a vodka martini, shaken not stirred. And whereas Bond is gambling in *Casino Royale*, *Thunderball* and *Goldeneye*, Fleming's agent, Ralph Izzard, whom he ran in Brazil found himself playing poker against his opposite number in German Navy Intelligence at a casino in Pernambuco. That episode also found its way into the Bond novels. Art imitated life.

A Boeing 707 jet airliner appeared again in the next Bond film, *From Russia with Love* (1963). From then on, jet airliners were associated with a glamourous, aspirational lifestyle.

And that, of course, was the whole point. Pan Am had to fill 50 Boeing 707s, and a bunch of celebs was never going to fill all those seats. No, Juan Trippe of Pan Am had bigger ideas. He was going to persuade middle America – predominantly white people with average incomes – to visit Europe. At that time, the wealth gap was narrower and the influence of the Jet Set on the general population was proportionately greater.

Trippe had a plan, and he followed it. In 1948, 750,000 passengers had crossed the Atlantic, half a million of them in ships, taking around five days. By 1957, the number of air passengers had doubled, and the number of ship passengers had declined to about the same as those flying overhead. Trippe started offering deals such as a fly-now-pay-later plan, and soon he had another 100,000 passengers. The Boeing 707 and DC-8 multiplied these numbers four-fold. The growth rate was so fast that that the 707 started looking small and Boeing started thinking about another number: 747.

Post-war Europe was dirt cheap, and if Britain was now a bit shabby, then other countries offered something of the Jet Set dream, in particular Italy. English-speaking Americans knew there was something special about Italy.

The Grand Tour had been a trip through Europe, a rite of passage taken by upper-class young Englishmen (and a few women) from around the 1660s to the 1840s. For England, an insignificant little country aground off the coast of Europe, it held huge cultural significance. Not only did the returning students bring back the dining fork and the umbrella, but they also returned with enormous enthusiasm for the art, culture and the origins of Western civilisation. With nearly unlimited funds, they commissioned paintings, bought ancient sculpture, drank too much and hired prostitutes whenever they could get away from their weary tutors, or 'bear-leaders', who were hired to keep them out of trouble. In short, the Grand Tour laid the foundations for the love of Greco-Roman antiquity and for the Renaissance, which endures today in most English-speaking countries.

Middle America was ready for the Grand Tour. Clutching their *Fielding's Guide to Europe*, they eagerly waddled out into the sunlit Roman piazzas.

Jackie Kennedy's year in Paris as a student had left her obsessed by all things French and chic. When she and her husband, John F. Kennedy, made a state visit to France in June 1961, she stood at the top of the steps of the Air Force One Boeing 707 and the world fell in love with her. Wearing a lemon-yellow Chanel suit with matching pill box hat, she descended and addressed the president, General de Gaulle, in perfect French. Later he remarked, 'She knows more French history than any French woman.'

For those who admired the court of Camelot, it would surely be a fine thing to visit Paris. And so in May 1962, it was easy for the Atlanta Art Association to fill all the seats of an Air France

Boeing 707 for a tour to France. The jet airliner was named *Château d'Amboise* (Air France's 707s were all named after châteaux of the Loire). Free champagne flowed, the Louvre was visited, art dealers were ransacked and the entire tour declared a huge success.

The return flight was designated Air France 007, and the chartered aircraft was another Boeing 707 named *Château de Sully*. Before the days of excess baggage, the 707[4] was laden with presents for those left at home, and the 122 happy Atlanteans looked forward to seeing them. The 707 pushed back from the terminal at Paris's Orly Airport at 1100hrs on Sunday 3 June 1962 and lined up on Runway 08. It was cleared for take-off at 1132hrs. With a piercing shriek from the four Pratt & Whitney JT4A-9 turbojets, the brakes were released and the aircraft set off down the runway.

After 48 seconds from the beginning of the take-off roll, the 707 reached the rotation speed VR of 158 knots and the pilot pulled back on the control column. According to witnesses, the nose was slightly raised but the aircraft failed to take off. After six seconds, the pilot dropped the nose and applied the brakes hard. But the aircraft had reached a speed of 179 knots and the runway was too short to stop in time. The accident report reads:

> After 250 m of braking the aircraft veered slightly to the left, and 50deg of flap were selected. Then after another 250 m the aircraft listed heavily to starboard. Its path then curved right, which suggests a possible attempt to ground loop[5]. However, the aircraft's speed precluded the success of this manoeuvre, and it left the end of runway while still on the centre line. It rolled for a while on the grass extension of the runway but, because of the unevenness of the terrain and the high speed of the aircraft (160 knots), the port

---

4 F-BHSM.

5 A ground loop is a rapid rotation of a fixed-wing aircraft in the horizontal plane (yawing) while on the ground. It can be used to produce a sudden stop, like a rally car's handbrake turn.

gear broke off 110 m from the end of the runway and was wrenched away. The aircraft pivoted left, and engines No. 1 and 2 scraped the ground. Fire broke out in the port wing at the level of the landing gear. About 300 m beyond the end of the runway the aircraft crossed the encircling road. The starboard gear collapsed, and No. 2 engine broke loose. It then struck the approach lights, which represented a considerable obstacle. It started to disintegrate when reaching the hollow at the end of the runway extension, which descends at a steep angle towards the Seine. The front part of the fuselage struck a house and garage. The nose of the aircraft broke away, and the rest of the fuselage came to a stop 100 m further on. The site of the accident was 550 m beyond the end of runway 08 on its extended centre line, at an elevation of 89 m.

Twenty thousand gallons of aviation fuel then blew up in a vast explosion.

One hundred and thirty people died. Two stewardesses and a steward, sitting in the tail of the aircraft were thrown out of the wreckage. One died later, one walked away. It was, at the time, the worst-ever single-plane disaster.

The air accident report cites two probable causes:

The accident was due to the concurrence of: 1) a considerable out-of-trim condition producing major loads on the control column at VR and VLOF [lift-off speed] which may have seemed prohibitive to the pilot-in-command; and 2) a failure of the trim servo motor control system which prevented the pilot-in-command from rectifying the faulty setting of the stabilizer and, consequently, from reducing the reaction at the control column. These factors led the pilot-in-command to discontinue take-off, but it was too late to stop the aircraft on the runway or slow it down sufficiently before the end of the runway.

All those antique paintings, perfume and the bottles of cognac bought in duty-free shops in the terminal may have contributed to the 'considerable out-of-trim'. The pilots might have been expected to know how heavy the aircraft was getting. The stabilizer failure was down to the aircraft and the maintenance regime.

Then, just 19 days later, there was another Air France Boeing 707 crash. Air France Flight 117 crashed into a hill while attempting to land at Pointe-à-Pitre, Guadeloupe, in the eastern Caribbean Sea. All 113 aboard were killed in the crash. Weather conditions were blamed.

Temple Fielding, writer of guidebooks and another nervous air passenger, had not approved of Air France, dubbing it 'Air Chance'. He had noticed that 'on every flight I've taken with this line, at least one tray of champagne or still wine or cognac has gone up to the cockpit'. Fielding asked the Air France boss, Max Hymans, why French pilots were free to drink during flight, while American pilots were obliged to abstain from alcohol 12 hours before take-off. 'He assured me that because the French pilot has grown up with wine,' Fielding reported, '"a little wine" won't hurt him during the flight.' Fielding was not satisfied: 'If Air France sincerely believes that the reflexes of their crews, after a glass or two of brandy or wine, are sufficiently razor-sharp to cope with instantaneous emergencies aloft, that's their affair... I regret that it's not the line for me.'

Alcohol was not mentioned in either accident report.

A total of 435 passengers died in four Boeing 707 crashes during nine months of 1962. In the following year, 124 passengers died in Boeing 707 crashes. In comparison, 67 passengers died in three Comet crashes in the 13 months between 3 March 1953 and 8 April 1954. Why was the Comet's reputation ruined and not that of the Boeing? Why did passengers keep buying tickets?

It could be that by 1962, jet aircraft were more accepted and recognised as safer than piston airliners. It could be that the British were more prepared to accept there was a design fault, commission a full inquiry and declare *mea culpa*. Or it could be that Boeing were more commercially minded and more determined to plough on. The answer is lost in the swirling mists of time, politics and commercial interests.

Come fly with me, let's take off in the blue...

One day after the Orly Boeing 707 crash, Frank Sinatra, performing at the Paris Lido, dropped 'Come Fly with Me' from the playlist. It was his standard show closer. Always nervous about flying with airlines, he chartered his own 707 when he wanted one, and chose his own pilots. In 1964, he bought one of the new Lear business jets, and in 1972, his own 707. Then, in 1977, he lent his Lear jet to his beloved mother, Dolly, for a 20-minute flight to Las Vegas. The weather was cloudy, then it became a blizzard, then the Lear flew into Mount San Gorgonio, killing Dolly, her friend and the Lear's crew.

Sinatra was utterly devastated. 'I could understand if it happened to me,' he mumbled to reporters.

The pallbearers at Dolly Sinatra's funeral included Frank's old buddies Jimmy Van Heusen, Leo Durocher and Dean Martin, whose own son was to die in another plane crash on Mount San Gorgonio, just a few years later.

Despite the carnage of the air crashes, the ironies and the ambiguities, the new Age of the Jet Set had begun. It was sung to Sinatra's seductive tune:

> *... Down to Acapulco Bay*
> *It is perfect for a flying honeymoon, they say*
> *Come fly with me, let's fly, let's fly away...*

Chapter 11

Fly Me

So far, women hadn't had much of a look-in during the Jet Age. There had been famous women pilots before the war, of course. There was Elinor Smith, the 'Flying Flapper of Freeport' who, for a dare, in October 1928, flew under all four of New York City's East River bridges, dodging ships as she went. There was Amelia Earhart, the first woman to fly across the Atlantic, who disappeared in July 1937 trying to fly across the Pacific. In Australia, there was Nancy-Bird Walton, who flew an air ambulance service in the Outback. And then there was the English-born Beryl Markham: aviator, adventurer, racehorse trainer and author. She was the first person to fly solo non-stop across the Atlantic from Britain to North America. She was also a member of the notorious Happy Valley set in Kenya, immortalised in the film *White Mischief*.

If you weren't exotic, rich and slightly mad, you were unlikely to join this select bunch of women.

Although British and American women weren't allowed to fly into combat during the war, the Soviets fielded units of all-women night bomber pilots in their teens and early 20s. They flew the Polikarpov Po-2 biplane, an aircraft that was so slow it was virtually impossible to shoot down: its cruise speed was 65mph (105km/h), far below the stall speed of the German fighters. The women flew at night, bombing behind the lines. They would throttle back and glide into the attack, and only the wind whistling in their biplane's wires heralded their arrival. The German soldiers likened the sound to broomsticks and called the women pilots *die Nachthexen* (the Night Witches).

During the war, 166 women pilots of Air Transport Auxiliary (ATA) flew aircraft around Britain between factories, maintenance units and airfields. They attracted enormous attention in the popular press. These women received the same pay as men of equal rank in the ATA, and this was a milestone as it was the first time that the British government gave equal pay for equal work.

It helped to be wealthy and upper class. Diana Barnato had been a debutant and was presented to King Edward. Her grandfather had been the immensely wealthy Barney Barnato of the De Beers mining company, and her father was Woolf Barnato, chairman of Bentley Motors. She became fascinated by aeroplanes when young and got her pilot's licence at the age of 20 at the Brooklands Flying Club in Surrey. In 1939, she was given a Bentley motor car for her 21st birthday. She applied to the ATA in 1941, was accepted and eventually delivered 260 Spitfires and hundreds of Hurricanes, Mustangs and Tempests. With more training, she converted to twin-engined types such as Mosquitos, Blenheims and Wellingtons, usually flying solo.

Then a new opportunity arose for women from ordinary backgrounds who wanted to get into the air. One day in 1929, a trained pilot and nurse named Ellen Church saw an advertisement on a San Francisco office window for the Boeing Air Transport service (BAT). She went in, and although the office manager wouldn't take her on as a pilot, she persuaded him that nurses ought to be employed on passenger aircraft to calm the public's fear of flying. Her suggestion was taken up by the BAT board and advertisements were placed. The 'sky girls', as BAT called them, had to be registered nurses, single, younger than 25 years old, weigh less than 115lb (52kg) and stand less than 5ft 4in (1.63m) tall. As well as attending to the passengers, they had to carry baggage, fuel the aircraft and help the pilots to push the aircraft into hangars. The salary was good at US$125 a month ($2,000 today).

Ellen Church wrote a manual, *Do's and Don'ts*, for stewardesses: 'A rigid military salute will be rendered the captain and co-pilot as they go aboard and de-plane before the passengers,' she instructed. 'Check with the pilots regarding their personal luggage and place it onboard promptly.' Stewardesses would have to address each passenger personally by name, serve tea, coffee and bouillon and she would have to carry a screwdriver, wrench and railway timetable in her pockets. This was because if the plane was forced down, she would have to redirect her passengers to the nearest railway station.

Ellen Church became the first stewardess to fly when, in 1930, she worked on a 20-hour flight from San Francisco to Chicago with 13 stops and 14 passengers. The new flying nurses would prove to be popular with nervous passengers. The new role of stewardess was entirely Church's idea, and she was therefore responsible for a huge opportunity for women. Ironically, after surviving countless flights and a world war, she died falling off a horse.

There had been other attempts to get women into the air. In the same year that Ellen Church had her brainwave, the Colonial Air Transport company ran a promotional offer. Wives could fly free on their flights, they proclaimed. After a huge number of offers were taken up, Colonial sent letters to their customers' homes, thanking them for their business and hoping they would see them both again soon. The Colonial offices were soon deluged with letters from angry wives saying they had never been on an airplane in their lives and what the hell was going on? But a number of secretaries had enjoyed the trip, thank you very much.

The other airlines swiftly took up Church's idea and by the 1940s, stewardesses were glamourous sky queens. 'Image was so important back then, and the job was so exciting,' remembered Helen McLaughlin, a Continental stewardess. 'Parents used to take their families on Sunday drives to the airport and they would wait to see the hostess come out of the plane. We were always told to wave!'

When the Jet Set arrived, the stewardess became a goddess: an icon of femininity. Passengers studied her face, her make-up, her complexion, her figure, her weight, her legs, her grooming, her nails and her hair. Most airlines would only hire white, unmarried young women, and they would have to resign if they married and retire at 32. Ruth Carol Taylor became the first African American airline flight attendant in the United States when she joined Mohawk Airlines in 1958, but she fell afoul of the no-marriage rule within six months.

In the 1960s, Delta Airlines acknowledged that flight attendants had other duties: 'In her new chic outfit, she looks like anything but a stewardess working. But work she does. Hard, too. And you hardly know it.' In fact, the women had to serve meals and drinks, calm nervous passengers and make sure safety rules were followed.

As the '60s ended, the hostess became the 'trolley dolly'. 'Sex sells,' insisted National Airlines. 'I'm Judy. Fly me…' its advert suggested. In its filmed commercial, a blonde woman removes her outer clothing and informs us, 'You can fly me morning, afternoon or night, just say when!'[1]

National claimed a 23 per cent increase in passenger numbers in the year following the campaign. Its next 'Take Me, I'm Yours!' campaign was inadvertently successful because National Airlines itself was taken over by Pan Am in 1980.

'*Coffee, Tea, or Me?*' chirped the *Uninhibited Memoirs of Two Airline Stewardesses*[2], but the work was more hard work than glamour. The two 'stews' described fighting off unwelcome attentions from pilots, and persuading passengers that the overhead luggage racks were not

---

1 https://www.flyertalk.com/articles/the-most-iconic-aviation-advertising-campaigns.html

2 Baker, Trudy and Jones, Rachel, *Coffee, Tea, or Me? The Uninhibited Memoirs of Two Airline Stewardesses*, Penguin, London (2003)

in fact their upper berth. With increased passenger numbers, the work became harder, and the pay became worse.

There was a backlash, of course. Flight attendants unionised in the late 1940s and '50s and demanded respect as safety workers. In the 1960s, they were among the first groups to take advantage of new laws prohibiting sex discrimination, and they challenged the airlines' restrictive employment policies and exploitive marketing practices such as figure-revealing uniforms. In this way, they were at the forefront of a new battle for women's rights.

Chapter 12

V Force: Britain's Cold War Jet Bombers

The American atom bombs dropped on Japan at the end of the Second World War changed warfare forever. The apocalypse visited upon Hiroshima and Nagasaki shook the world: one device dropped from one aircraft could now equal all the TNT that rained down on Germany in the Second World War. Other countries began seeking ways of making their own nuclear weapons and the means to deliver them.

This had all been foreseen by H. G. Wells, of course. In his novel *The World Set Free*, written before the First World War in 1913, Wells prophesied that 'a man could carry about in a handbag an amount of latent energy sufficient to wreck half a city'. He also described a nuclear war conducted by bomber aircraft. He even named the devices 'atomic bombs'. His knowledge came from reading work by the father of nuclear physics, Ernest Rutherford, and Frederick Soddy, who discovered the disintegration of uranium.

It is possible that Wells' book even influenced the invention of nuclear weapons. The Hungarian American physicist Leó Szilárd read *The World Set Free* in 1932, the same year the neutron was discovered. And in 1933, Szilárd conceived the idea of neutron chain reaction and filed for patents on it in 1934: 'Knowing what it would mean – and I knew because I had read H. G. Wells – I did not want this patent to become public.'[1]

The British had first demonstrated the nuclear nature of the atom, discovered the neutron, then split the atom. The discovery of fission raised the possibility of the release of vast quantities of energy. They were, therefore, in the position of being able to make the first 'atomic bomb'. Working at Birmingham University, the German refugee physicists Rudi Peierles and Otto Frish co-authored a paper showing how to construct an atomic bomb from a small amount of fissile uranium-235.

The MAUD Committee, meeting at the Royal Society in Burlington House in London in April 1940, explored the feasibility of fission weapons (a scientist's housekeeper provided the code name[2]). It concluded that 25lb (11kg) of uranium could produce the explosive power of many thousands of tons of TNT. Britain began a nuclear weapons project, codenamed Tube Alloys. Materials were hastily secured: heavy water was rescued from France during the German invasion and initially kept at HM Prison Wormwood Scrubs but was later stored at Windsor Castle alongside the Crown jewels. Nazi Germany was also racing to build the bomb, so, lacking resources, the British sent 19 scientists to help the US with their Manhattan Project, which exploded the first atom bomb in 1945.

However, after the war, the Americans refused to share their knowledge with the British. The US Atomic Energy Act of 1946 (McMahon Act) prohibited the export of atomic knowledge, even to countries that had collaborated on the Manhattan Project.

---

1 MacKenzie, N. and MacKenzie, J., *H.G. Wells: A Biography*, Simon & Schuster, New York (1973), p. 299

2 The name MAUD came about in an odd way. On 9 April 1940, the day Germany invaded Denmark, Niels Bohr had sent a telegram to Frisch. It ended 'Tell Cockcroft and Maud Ray Kent'. Codebreakers thought it referred to radium, hidden in an anagram. They replaced the 'y' with an 'i', producing 'radium taken'. When Bohr got to England in 1943, he told them the message was addressed to John Cockroft and Bohr's housekeeper, Maud Ray, who was from Kent. And so, the committee was named the MAUD Committee.

In 1945, the new Labour government debated whether or not to build Britain's own nuclear deterrent. It was feared that the Americans might withdraw from Europe, taking with them the nuclear umbrella. 'We've got to have this thing over here, whatever it costs,' left-winger Ernest Bevin told one committee, 'and we've got to have the bloody Union Jack on top of it.'[3] It was pointed out that a nuclear deterrent was a cheaper option than huge standing armed forces, and it was believed that the concept of mutually assured destruction would inhibit the use of atomic weapons. Britain decided to go it alone.

Then, in 1949, the Soviet Union exploded their own bomb: 'Joe One'. It emerged that Klaus Fuchs, a scientist working on the British and American bombs, was a Soviet spy and had passed on vital secrets. So had others. Once again, the Americans refused any sharing of research: the British clearly could not be trusted with nuclear information. As a result, in 1950 a centre for advanced nuclear research was created at Aldermaston, and a nuclear reactor to produce plutonium was built at Windscale.

The transporting of the plutonium was extraordinarily amateurish: it was loaded into lead containers and put in the back of a Vauxhall estate car to be driven to Woolwich for further testing. On the way, this car broke down outside a pub in South London, leaving Britain's nuclear deterrent sitting on the side of the road.

Aircraft to carry the new bomb were required. They would use jet engines and, it was hoped, be able to fly higher than Soviet fighters. And so, in November 1946, the Air Ministry issued the operational

---

3 Unexpectedly, the Trade Unionist Bevan was something of an imperialist. He also was a member of the Garrick Club and ran a flamboyantly yellow Talbot Darracq motor car. As Foreign Secretary, he used his brusque Trade Union manner to deal effectively with Stalin. It was said there was only two posts in the Foreign Office he could have held: Foreign Secretary or doorkeeper.

requirement OR230 for an advanced jet bomber capable of carrying a 10,000lb (4,500kg) bomb to a target 2,000 nautical miles (3,700km) from a base anywhere in the world with a cruising speed of 500 knots (930km/h) and at an altitude of between 35,000ft and 50,000ft. A later version of this requirement dictated the use of shorter runways.

It was clear that much more powerful and economical jet engines would be needed, and so the Ministry turned to the engineers and asked, 'Can you do this?'

It had been done before. Once the piston aero engine had been established, it became the subject of intense development. The Wright's 1903 'Flyer' four-cylinder engine of just 12hp led to the Pratt & Whitney R-4360 Wasp Major 28-cylinder four-row radial piston engine of no less than 71 litres and 3,500hp. And all this in just 40 years.

The jet engine eventually would follow a similar trajectory, with the pioneer Whittle engine of 480lb of thrust leading to the General Electric GEX9 engine for today's Boeing 777X developing 134,300lb of thrust. Oddly enough, in both cases the increase in power is in the order of 300 times.

Readers of my *Merlin* book will know how the piston engine was persuaded to make more power: by being made to swallow more air and fuel. The jet engine was no different.

Although Frank Whittle was justifiably proud that his engine had only one major moving part, further complication was inevitable as soon as other engineers brought their minds to the problem of making more power.

Unlike its competitor, Rolls-Royce, the Bristol Aeroplane Company had always made its own aircraft as well as its aero engines. It had concentrated on air-cooled radial piston engines during the 1930s, and its enormous 39-litre 14 cylindered two-row sleeve-valved

Hercules[4] engine had consistently shaded the Rolls-Royce Merlin engine in power output: the 1939 Hercules I engine gave 1,290hp; this soon improved to 1,375hp in the Hercules II. The main version was the Hercules VI, which delivered 1,650hp, and the late-war Hercules XVII produced 1,735hp. These were big engines with a large frontal area, so they were more suitable for bombers or large fighters like their Bristol Beaufighter. Many British frontline aircraft were powered by Bristol air-cooled radial engines.

As a result, the Bristol Engine division was keen to beat Rolls-Royce in the new field of jet engines, and so immediately after the end of the war it turned its mind to the job, concentrating on increasing thrust and decreasing fuel consumption. Operational Requirement OR230 gave it a target to aim for: the engines for a nuclear bomber.

Bristol realised that the early jet engines with one shaft or spool were a compromise, as the airflow conditions at the front of the compressor were quite different to those at the back of the compressor. So it came up with the idea of a two-spool engine, where a high-pressure turbine is driven by the combustion gases, which, in turn, drives a high-pressure compressor, and a low-pressure turbine similarly drives a low-pressure compressor. A clever feature was that the low-pressure turbine spun on its own separate shaft, which passed through the centre of the high-pressure shaft. This two-spool jet engine design is essentially an engine inside an engine. The advantage is that each compressor runs independently at its best efficiency. The low-pressure shaft might revolve at 6,000rpm and the high-pressure shaft might turn at 10,000rpm.

There is an analogy with modern sequentially turbocharged car engines, where one small and one large turbocharger cover the rpm range of the engine, which uses one turbocharger for lower engine speeds and a second or both turbochargers at higher engine speeds.

---

4 Bristol engines were named after ancient Greek mythological characters.

With two spools the jet engine's fuel consumption is lower than with a single-shaft engine over a wider range of operating conditions. This would result in a greater range. There could be more than two spools: later Rolls-Royce engines would feature three spools.

The Bristol Olympus was the world's first two-spool axial-flow turbojet engine design, dating from November 1946, and it first ran in 1950. It was intended for the Avro Vulcan bomber and was also used in the supersonic airliner Concorde. A marine version was used to power warships and land-based versions still provide quick start-up peak demand power for the National Grid in electrical power stations. Hundreds of Olympus generator sets now power oil rigs around the world.

Wartime experience with the Avro Manchester and Lancaster showed that planners could not tell which aircraft design would prove successful, so three jet bomber designs were chosen: two extreme versions and one 'banker'. They would be known as the V-Force. The Handley Page Victor and the Avro Vulcan looked like spacecraft. The Vickers Valiant was the more conservative design and was the first to fly. Of the three V bombers produced, the Vulcan was considered the most technically advanced and hence the riskiest option.

On 2 October 1952, during Operation *Hurricane*, HMS *Plym*, a small frigate that had served during the war, was deliberately blown up in the first British atom bomb test in the Monte Bello Islands in Western Australia. The yield was calculated partly by using oil paint lead tubes laid out near the bomb and calculating what pressure was needed to crush them. The result was 25 kilotons or 25,000 tons of TNT: the equivalent of 4,000 Avro Lancaster bomb loads. The re-elected prime minister, Winston Churchill, was cock-a-hoop: Britain now belonged to an exclusive club. Three weeks later, though, the US exploded 'Ivy Mike', the first hydrogen bomb, 400 times more powerful than the British bomb. Then, a year later, the Soviets exploded their own hydrogen bomb: 'Joe Four'.

By the mid-1950s, there was public alarm at the number of bomb tests and a 'Ban-the-Bomb' movement grew. By then, the Vickers Valiant bomber was ready to fly. It was using four of the Rolls-Royce Avon turbojets, with 10,000lbf (44kN) thrust and specific fuel consumption of 0.932lb/(lbf·h). On 11 October 1956, a Valiant was the first RAF aircraft to drop a British operational atomic bomb when it performed a test drop of a Blue Danube weapon on Maralinga, South Australia.

Public disquiet grew, and so in Operation *Grapple X*, Britain raced to test its own hydrogen bomb before a ban on atmospheric testing. On 15 May 1957, a Valiant dropped the first British hydrogen bomb, the Short Granite, over the South Pacific.

The US did not like the idea of Britain having an independent nuclear deterrent; former United States Secretary of State, Dean Acheson, said:

> Britain's attempt to play a separate power role – that is, a role apart from Europe, a role based on a 'special relationship' with the United States, a role based on being the head of a Commonwealth which has no political structure or unity or strength and enjoys a fragile and precarious economic relationship – this role is about played out.

He added that Britain had lost an empire but had not found a role. He went on: 'Great Britain, attempting to work alone and to be a broker between the United States and Russia, has seemed to conduct a policy as weak as its military power.'[5]

However, the V-Force gave Britain independence from the US. The Valiant aircraft was able to be used during the Suez Crisis of 1956 when the US refused to support British actions. The same situation

---

5 https://www.theguardian.com/century/1960-1969/Story/0,,105633,00.html

arose during the Falklands War when Vulcans were used when no overt military support was forthcoming from the US.

As for the much trumpeted 'special relationship' between the US and the UK, it looked more and more like the kind of relationship Oedipus had with his parents.

The V-Force bombers were designed to drop their nuclear bombs from high altitude, with the assumption that the Soviets could not reach them with fighter aircraft or missiles. They carried no defensive weaponry, relying instead upon high-speed high-altitude flight to evade interception. This was a dangerous assumption to make, and it was to prove an expensive mistake.

When, in 1960, the Lockheed U-2 of Gary Powers was shot down by a Soviet missile at around 70,000ft, the V-Force had to change their tactics to low-level attack, hoping to fly under the Soviet radar. This new turbulent altitude had a disastrous effect on the Vickers Valiant airframe. Wing spar attachment forgings started cracking due to the use of an aluminium alloy, DTD 683. This had been developed during wartime for aircraft whose lives were necessarily short. It was found that this alloy was crack sensitive and prone to failure after many repeated cycles. After more failures, the entire Valiant fleet was scrapped. When asked in 1965 to make a statement in the House of Commons, Secretary of State for Defence Denis Healey stated that it 'was not in any way connected with low-level flying'.

There was more to this than meets the eye, as the fatigue appeared to affect all Valiants, not just those used for low flying. It all seemed so horribly reminiscent of the Comet.

The Handley Page Victor was more successful than the Valiant. It was more streamlined, which made it look like a rocket ship, and it had a novel crescent-shaped wing that allowed it to break the sound barrier on several occasions.

A remarkable woman mathematician, the German-born Johanna Weber, had joined the aerodynamics department at the RAE in Farnborough and had worked on linear and simple aerodynamic models, which were calculated by hand by a team of women 'computors'. She contributed much to the exceptional aerodynamics of the Handley Page Victor and would go on to a crucial discovery that enabled the building of Concorde.

The Victor had a best range of 6,000 miles (9,650km) and a greater maximum bomb load of 35,000lb (15,900kg). It mounted four Armstrong Siddeley Sapphire turbojets with a thrust of 11,050lbf (49.2kN) and a specific fuel consumption of 0.885lb/hr/lb (90.214kg/kN/hr). It also carried a small turbo-shaft gas turbine to act as an APU.

Due to its high-mounted tail, the Victor had one endearing characteristic much enjoyed by prank-prone pilots: it could land itself. Once lined up on the runway and the engine power reduced, the pilot could sit back with his arms folded. The wings would enter into ground effect, but the tail would continue to sink slowly, so the aircraft gently settled onto its wheels at the correct angle. Once again, though, low-level flying resulted in fatigue cracks and most Victors were converted into fuel tankers for jet fighters.

The delta-winged Avro Vulcan was the most futuristic design and was the most manoeuvrable of the rival strategic British bombers. Beside it, the Boeing B52 bomber looked like a flying factory. The Vulcan's barrel roll at the 1955 Farnborough Air Show demonstrated that this heavy bomber was something unusual. The delta wing also inadvertently made the Vulcan the first stealth nuclear bomber, as, on radar screens, it sometimes disappeared altogether.[6] This was because the thick wing, buried engines, vestigial fuselage and small vertical stabiliser provided few radar returns. It also possessed advanced electronic countermeasures (ECM), so the British bomber was hard to detect. In contrast, the Boeing B-52, with its four double engine pods, slab-sided fuselage and huge tail was a giant radar reflector. The Soviet TU-95 *Bear* was even worse, adding four huge 8-bladed 20ft (6m) propellers that reflected every radar signal in sight.

As a result of their stealth profile and ECM, during the 1960s US Operation *Sky Shield* war games, Avro Vulcans successfully penetrated American defences, 'nuked' New York and, on one occasion, flew so low that a wingtip brushed the ground. They could also fly at 55,000ft at 646mph (1,040km/h), or Mach 0.96.

The Vulcan had brand-new Bristol Olympus twin-spool turbojet engines, with 11,000lbf (49kN) thrust each. Their specific fuel consumption was 0.809lb/(lbf·h) (22.9g/(kN·s).

One Vulcan accident illustrates the relatively slow throttle response of turbojet engines compared to that of piston engines. On 7 October 1964, a Vulcan was practising an asymmetric power practice approach at RAF Coningsby: two engines were producing thrust with two at idle. When the pilot started an overshoot of the runway, he moved all the throttles quickly to full power. The engines that had been producing power reached full power more quickly than the engines at idle and the resultant asymmetric thrust exceeded all the available rudder authority, causing the aircraft to spin in and crash. All the crew died. If only the pilot had known.

There was another 'if only' concerning the Vulcan: it could have been a jet airliner. In response to a Ministry of Supply requirement for a military and civilian aircraft for long-range duties, Avro proposed the Avro Atlantic, a civilian version with up to 113 passenger seats.

---

6 Sweetman, Bill, 'The Bomber that radar cannot see', *New Scientist* (4 Mar 1982)

Oddly enough, these faced backwards, a safety requirement that the RAF still insists on in its transport aircraft.

Passenger-carrying versions of the Vickers Valiant and Handley Page Victor V-bombers were also planned. The Vickers V-1000 won the contest. However, once again the Ministry of Supply then vetoed the idea, the reason given being the usual increase of weight during development.

In the House of Commons Paul Williams MP protested:

> The weight growth has been completely matched, through the years, by an equal growth in the engine power needed to get this aircraft into the air... on present calculations it will have an engine which will produce a greater thrust than the comparable American civil aircraft. While mentioning the engine it may be worth recalling that this aircraft has been deliberately designed around the Rolls-Royce Conway engine. Because of its design that engine is both more efficient and more economical than any other in the world, on Transatlantic routes as well as others.[7]

Once again, British bureaucrats had cancelled a viable project. Vickers' managing director, George Edwards, was despairing: 'We have handed to the Americans, without a struggle, the entire world market for big jet airliners.'

The Avro Vulcan was nearly scrapped without ever going into action. But then, in April 1982, just two months before the Vulcans had been due to retire, Argentina seized the Falkland Islands. The RAF planned seven missions named Operation *Black Buck*, the first two to be night-time attacks on Port Stanley airport, flying a 7,600-mile

---

7 *Hansard*, vol. 547 (8 December 1955)

round trip from Ascension Island. This was far beyond the Vulcan's range, so two Vulcans were to be refuelled by no less than 11 Victor K2 tankers (which also refuelled each other) to bomb the airport and other targets in the Falklands.

The Avro Vulcans were selected based upon their engines: only those with the more powerful Bristol Olympus 301 engines were chosen. The power of the Olympus had doubled by 1982. These had then been de-rated but were 'hotted up' for *Black Buck*, giving 21,000lbf (93kN) of thrust.

Operation *Black Buck* was a masterpiece of logistics, but the results were debatable: during both the *Black Buck 1* and *2* raids, 21 1,000lb (450kg) bombs were dropped across the runway of Port Stanley Airport for the expenditure of 400,000 gallons of fuel at a cost of £3.3 million. This did not close the runway to Argentine Hercules aircraft, but it probably stopped its use by fast jets.

Then, during *Black Buck 4, 5* and *6*, the Vulcans fired Shrike anti-radar missiles, acquired from the United States, destroying two small air defence radars, killing four Argentine personnel and lightly damaging the larger radar station that was their primary target. And during *Black Buck 7*, all 21 bombs missed their targets.

Assessments of *Black Buck* were not complimentary. It was pointed out by a Royal Naval commander that, for the same amount of fuel burned on the first mission, the Sea Harriers of the carrier force could have carried out 785 sorties that would have delivered 2,357 bombs. And an American Marine Corps study concluded that the RAF was seeking a piece of the action, and that the results were not worth the effort.

It might seem odd to us now that three different makes of jet engine were used in three similar aircraft with no thought of standardisation, but just after the war, there was a multitude of British aircraft manufacturers. Inevitably, these would shake out into just a few.

However, the V-Force bombers did share several features: each was intended to drop a single nuclear bomb. Each mounted four jet engines. Each had five crew: pilot, co-pilot, two navigators and an air electronic officer, although only the first two crewmembers had ejection seats – a 'mixed' escape system. As a result, there were several occasions when only the pilots managed to eject safely, and this caused a certain amount of disquiet and poor morale from those only issued with parachutes.[8] And at first, all V-Force bombers were painted with white 'anti-flash' paint designed to reflect the emissions from a nuclear blast. When they switched to the low-altitude role, they were painted with camouflage.

The new atom bombs were lighter than the old TNT bombs and tended to fly back up when released, sometimes climbing back into the bomb bay, a disturbing state of affairs. A series of 'Dragon's Teeth' spoilers at the front of the bomb bay solved the problem, sucking the nose of the nuclear bomb downwards in the right direction.

By the time the V-Force reached 180 aircraft, the concept behind high-altitude bombers was outdated. Because of the danger to the bombers at low level, standoff weapons were introduced. These would allow the V bomber to release a self-propelled nuclear missile 100 miles away from the target, thereby avoiding the local defences. This programme started with the Blue Steel missile, and it was then planned to move to the much longer-ranged American Skybolt air-launched ballistic missile. When John F. Kennedy cancelled Skybolt, the survivability of the V-Force crews was in doubt: the new Soviet defences would ensure these bombers wouldn't get through.

It was the end of the V-Force.

---

8 Blackman, Tony and Wright, Anthony, *V Force Boys: All New Reminiscences by Air and Ground Crews Operating the Vulcan, Victor and Valiant in the Cold War and Beyond*, Grub Street, London (2017)

This eventually led to the Royal Navy taking over the nuclear deterrent role from 1968, using UGM-27 Polaris intercontinental ballistic missiles launched from nuclear submarines.

The 1962 Cuban Missile Crisis put Britain on high alert and the V-Force was on a four-minute standby, bombed up and ready to attack Moscow. Two jets in every major RAF base were armed with nuclear weapons and were on permanent standby called the Quick Reaction Alert. My brother remembers seeing Avro Vulcans standing on the tarmac at RAF Cottesmore, Olympus engines running and with the pilots' heads visible.

Later, in the 1970s, a microwave cooker in the next village to ours nearly started the Third World War. The frontline radar station at RAF North Luffenham went on Red Alert when a massed formation of Soviet nuclear bombers suddenly appeared all over the screens. A newly installed oven in the nearby Horse and Panniers pub was the source of the microwaves.

As a footnote to the British nuclear deterrent, it must be remembered that not all atom bombs were delivered by air. Some were powered by chickens. Blue Peacock was a 10-kiloton nuclear mine designed to be buried in Germany in the path of invading Soviet tanks. However, in the winter these became too cold, and their electronics stopped working. It was decided that live chickens, supplied with food and water could be sealed inside the bomb. Their body heat would be sufficient to keep the mines' internal components at the correct temperature. When this plan was revealed to the media on 1 April 2004, there was a general outcry against a particularly cruel joke. This received the frosty reply, 'It does seem like an April Fool, but it most certainly is not. The Civil Service does not do jokes.'[9]

If the British Isles had been incinerated by a Soviet strike, the V-Force bombers were instructed not to return. At sea, the UK's nuclear submarine captains had to listen out for BBC Radio 4's *Today* programme. If it was not broadcast for three days, the UK was deemed to have been destroyed, and the Prime Minister's Letter of Last Resort opened. This could give permission for the submariners to fire their nuclear missiles into Soviet Russia. This would ensure mutual destruction.

In retrospect, the decision to build three different versions of nuclear bomber now seems foolishly extravagant. Once it had become clear that the ambitious Vulcan was viable, surely the Victor and the Valiant should have been cancelled? And what had the V-Force jet bombers achieved for their vast cost? Only some damage to three Egyptian airfields during the 1956 Suez Crisis and the debatable results of the 1982 Operation *Black Buck*.

Plus, the incalculable value of deterrence.

Mercifully, they never had to be used for their intended role, which would have resulted in the horrific death of hundreds of thousands of Russian civilians.

Perhaps it was that knowledge of what the V-Force could do that preserved the *Pax Atomica*.[10]

---

9 http://news.bbc.co.uk/1/hi/uk/3588465.stm

10 *Pax Atomica*: 'Atomic Peace' derives from *Pax Romana*, which describes the period of stability during the Roman Age.

Chapter 13

Lightning Conductor

The V-Force jet bombers would have been the first targets of a Soviet air attack on the UK. To defend them, a supersonic British jet fighter was developed after a 1947 proposal by the Westland chief designer, 'Teddy' Petter. The English Electric Lightning was a brutally powerful jet fighter employing two Rolls-Royce Avon turbojet engines stacked vertically. It was designed to climb like a rocket, intercept and then destroy Soviet nuclear bombers. The priorities were acceleration, rate of climb and top speed. Range, often a weakness of British fighters, was just enough to allow a radius of operation of 150 miles (240km) from the V-Force bomber airfields. Due to the afterburner, after take-off the Lightning could climb vertically and ultimately as high as 88,000ft.

On 22 July 1966, an RAF engineer accidentally engaged the afterburner of a Lightning during ground testing. Unable to disengage it, he shot down the runway, narrowly missing a fuel bowser and then a de Havilland Comet, which was taking off, before taking off himself. Flying without a helmet or canopy, the ejection seat disabled, and the landing gear locked down, he found himself in a bit of a pickle. The story has become something of an aviation myth, so it is worth hearing the story from the unwilling pilot in his own words.

The Lightning was a complex aircraft, and this, the second example built, had one particular problem on that July morning in 1966, on the runway at RAF Lyneham: 'XM 135 was being prepared for despatch to a Target Facilities Flight, but over a period of weeks it had been giving no end of trouble.' The speaker was an RAF engineering officer, Wing Commander 'Taffy' Holden. As the man in charge of maintenance of the RAF jets at Lyneham, he just had to get this particular aircraft away:

> I had Canberra, Meteor and Lighting types which were gradually being prepared for dispatch to various flying unit tasks. When the Meteor and Canberra types had been cleared, the powers that be decided that the MU (Maintenance Unit) should be closed after the last Lightning had been dispatched. Up until the last Canberra, I had a qualified and current test pilot on my staff for those aircraft, but he was not a current Lightning pilot. When a Lightning needed test flying, I had to call for any available pilot with a current test pilot rating.

The problem with XM 135 was electrical. Every time it was flight tested, the pilot found that during the first few yards of the take-off run the power to the primary flight instruments would cut out. The electricians on the base had tried everything they could think of, and now they needed another flight test.

There was no Lightning pilot available, so Taffy Holden decided to conduct a ground test himself. This would involve a simulated take-off run while switching in and out of various temporary circuits in an attempt to find the fault:

> I was correctly strapped into the cockpit (seated on the in-situ parachute and ejection seat) and after starting the engines and holding the aircraft static, on brakes, I did the necessary preliminaries for the electrical checks in the cockpit, checking the notes I had scribbled on a notepad which lay on the coaming in front of me.

Fortunately, Holden had learned to fly on Tiger Moths and had some experience on the piston-engined Chipmunks. He wasn't quite ready for a jet-engined fighter, though. And he certainly didn't know that if he pushed the throttles too far, they would lock into afterburner mode, or 'reheat'. To unlock reheat, one had to feel for an unlocking key.

Holding the brakes, I gradually opened the throttles to about 90%. My feeling at the time was the unexpected heavy vibration of Avon power held against the brakes. I did a quick check of the temporary electrical switches and circuitry lights, then released the brakes. The initial punch from the thrust was quite remarkable and I moved the expected 30 to 40 yards before I throttled back and applied the brakes. So far so good.

Air Traffic Control (ATC) was communicating by radio and was happy for him to make another short run:

ATC had also been holding up a fuel bowser and trailer with 3,600 gallons of AVTAG for a waiting C-130 aircraft refuelling; they decided to allow the bowser to cross the runway. On opening the throttles for the final test, I obviously pushed them too far, misinterpreting the thrust, because of the unexpected heavy vibration and they got locked into reheat. Yes, I did use some expletives, but I had no time to think about of getting out of reheat, because in front of me, the bowser and trailer had just crossed the runway, from left to right, so my thoughts were to make sure I was missing them by sufficient margin. No, I couldn't steer to clear them; reheat takes you in a straight path like a bullet out of a gun. The time between finding myself in reheat and just missing the bowser was less than half the time I have taken to write this sentence.

Before my thoughts could again return to getting myself out of reheat, I was gathering speed and about to cross the main duty runway, where a Comet had just passed on its take-off run. I then had no time to look for reheat gate keys, my eyes were on what next lay ahead. Two things, the end of the short runway 07 and just beyond was the small village of Bradenstoke, which I just had to miss. There was no chance of stopping, none whatsoever. I had gained flying speed (that is what reheat is for, short sharp take-offs) and I had no runway left. I did not need to heave it off the runway, the previous test pilot had had trimmed it exactly for take-off and only a slight backward touch on the stick and I was gathering height and speed. Then my thought was to get my speed back in case I should damage the undercarriage. Incidentally, I could not have raised the under-carriage: the ground locks were in place for safety reasons. With only clear blue sky in front of me, I could then search and feel for those gate keys. Yes, I found them and thanked my lucky stars that my engine foreman had quite incidentally told me of their location, and I was soon able to get my speed back to (I am guessing now) about 250 knots. My next thoughts were to keep Lyneham airfield in sight and where had the Comet got to, the one I had missed a few seconds ago! Then I asked myself, should I eject and where and when? No, I could not; the safety pins were in the ejection seat and safe for servicing, not flying. My only alternative then was to attempt a landing, but how does one interpolate or extrapolate Tiger Moth, Chipmunk flying to a two engined, 11-ton, beast like the Lightning?

After regaining my bearings, a little composure and simply by observation, making sure that the Comet had been warned away, I decided I should attempt a landing on the duty runway and direction. I was trying to combine all my limited flying experience into a few minutes of DIY flight 'training' on a Lightning. It wasn't easy, but I must admit that some of the elementary rudiments of my proper flying training and flight theory were coming in useful. I needed to get the feel of the aircraft if I was to get it back on the ground. My first approach was ridiculous, I could tell that my speed, height, rate of descent, even alignment wasn't correct, and my best plot was to go around again. This time making sure that my throttles would be well below reheat position. A second approach was no better, I had some aspects better, but as the duty runway was on the lip of an escarpment, with a valley floor beyond, my rate of descent took me below runway height, and I found myself adding power to get back to the right level. More power meant more speed and I was trying to get to something like 150 knots for landing, but the uncoordinated attempt was becoming a mess, so I abandoned it, took myself away on a very wide circuit of Lyneham and decided to land in the opposite direction. This I thought would give me more time to get the 'feel' right and if I made a mess of the landing, I would overrun the runway and just drop (crash) into the valley beyond. In that direction, with a messed-up landing, I would have no fear of crashing into Lyneham village.

The long final leg of this approach gave me the thinking time that I needed, and I gradually got the feel that speed, alignment, rate of descent, height and approach angle were better. I plonked it down at about the right position off the threshold, but just forgot that I was in a nose wheel aircraft and emulated my best three-wheelers in a Chipmunk or Harvard. The result was that I crunched the rubber block which encases the brake parachute cables. However, I had got down, but then I had to stop.

I obviously knew that the Lightning had a brake parachute, but where was the 'chute lever', button or knob? There, I found it marked Brake Chute and I pulled it and I could then look ahead and concentrate on keeping straight and somewhere near the centre of the line. I hung on the brake lever, I wasn't slowing as much as I would like, so I just kept up my hand pressure on the brakes. I had about 100 yards of runway left when I stopped and, even then, I didn't know that the brake parachute had dropped off as soon as it was deployed, because the cable had been severed as a result of my super tail wheel three-pointer.

The Lightning was towed into a hanger and Holden was given sedatives by the medical officer to calm his nerves. The only repairs needed on the Lightning were a new rubber block under the tail and some new brakes. For Holden, though, there was long-term damage from his terrifying experience: 'Some years after the incident, my hidden fears of high-speed flight came to the surface, and I had to spend two periods in hospital. I had not come to terms with the emotional side of the event.'[1]

As a postscript, the fault turned out to be a spare wire that was shorting out on the UHF radio when it moved back slightly on its trunnions during take-off runs...

Twin engines were a good idea for a high-speed jet fighter, and the Lightning used the most powerful Rolls-Royce Avons, developing 12,690lbf (56,450N) of thrust and (as Taffy Holden found out) 17,110lbf (72,770N) with reheat, or afterburning.

Stacking one Avon on top of the other might have been a good idea for frontal area reasons, considering that most pilots are taller than they are wide, even when sitting down. But any leaks of oil or fuel from the No. 2 top engine onto the No. 1 bottom engine would result in a fire. The engine casings glowed red hot at 600°C, and so gold paint had to be used to protect the fuselage.

Even at idle speed, the two Avons were powerful enough to drive the Lightning along the runway at 80mph (130km/h), about the top speed of a 1912 biplane, so engine No. 1 was usually shut down after landing to preserve the brakes. These were Dunlops Maxaret, the first anti-lock braking system (ABS), which is now so widely used on cars. They were developed for the Lightning and the V-Force bombers and then found their way onto the Jensen FF, the 1965 British sportscar that introduced ABS, all-wheel drive and a traction control system, many years before the car manufacturer Audi heavily advertised these features. Such is the value of the high-tech aero industry.

---

1 'The Wheel', NZ Ex-RAF Apprentices Newsletter (21 September 2013)

Chapter 14

Stratosaurus Rex: The 100-Year-Old Bomber

A nother brand-new twin-spool jet engine was revealed in 1950, this time by US aero engine manufacturer Pratt & Whitney. It was dubbed the JT57. It would be fitted onto the Boeing B-52 Stratofortress, a new Cold War bomber, together with seven of its siblings.

Wait, how many?

Yes, the gigantic B-52 has no less than eight jet engines. Furthermore, some B-52s will still be flying in the 2050s, 100 years after they were built, with the same wings, the same fuselage but not (we hope) the same pilots.

The B-52's predecessor, the Boeing B-47 Stratojet, was on the drawing boards in May 1945 when the chief of Boeing's technical staff, George S. Schairer, travelled to Germany to study captured swept-wing designs. After seeing models and wing-tunnel data, he telegraphed his office urgently: 'Stop the bomber design.' On his return, he initiated a 35° swept wing for the B-47. The result was a modern design with shoulder-mounted wings and with jet engines in pods. That design change pioneered the look of jet airliners for the next century.

Despite having the speed and manoeuvrability of the piston-engined fighters of the late 1940s, the B-47 suffered metal fatigue in the wings, just like the British Victor and Valiant of the same era. No less than 203 were lost in crashes, with 464 deaths, about 10 per cent of the total number produced. And with six General Electric J47-GE-25 turbojet engines, developing only 7,200lbf (32kN) thrust each, there wasn't really enough power.

The B-47 was also rather careless with its nuclear bombs, losing several in what the US delicately calls 'broken arrow' incidents. On 10 March 1956, a B-47 carrying two nuclear weapons from Florida to an overseas base just disappeared. It was supposed to rendezvous with a tanker over the Mediterranean but failed to make contact. No trace of the aircraft or its lethal load was ever found, and its location still remains a mystery.

Then, just 139 days later, on 27 July 1956, a B-47 crashed at RAF Lakenheath in Suffolk, England, killing its crew and careering into a storage igloo containing three Mark 6 nuclear bombs. There was an intense fire. Later, picking through the wreckage, bomb disposal officers were appalled to find that one of the Mark 6s had its detonators sheared off. It could so easily have exploded.

On 5 February 1958, a B-47 collided with an F-86 jet fighter near Savannah, Georgia. It was carrying a Mark 15 nuclear bomb, which it jettisoned in water just off Tybee Island. The aircraft landed safely, but the bomb was never recovered. Assistant Secretary of Defense W. J. Howard stated that the Tybee Island bomb was a 'complete weapon, a bomb with a nuclear capsule'. It is still there, somewhere.

Then again, in England just three weeks later, at the US base at Greenham Common, a B-47 developed problems shortly after take-off and jettisoned its two 1,700-gallon external fuel tanks. One hit a hangar while the other struck the ground 65ft (20m) behind another parked B-47. This bomber, which was fuelled, had a pilot on board and was reportedly carrying a 1.1-megaton B28 nuclear bomb. It immediately caught fire. This took 16 hours to extinguish, partly because the magnesium alloys used in the B-47 were prone to burning with an extremely hot incandescent flame. The US government has never confirmed whether this accident involved a nuclear warhead.

And two weeks after that, on 11 March 1958, a B-47 bombardier, who was checking a warning light, accidentally pulled out a pin holding in the Mark 6 nuclear bomb. It dropped, pushing open the bomb doors. Down on the ground at Mars Bluff, South Carolina, three girls were playing near a playhouse built by their father. The bomb hit the building and its high explosive blew up, injuring all three girls and three other members of their family. Luckily, the fissile core was stored elsewhere on the aircraft.

The B-52's 100-year-long career will span all the significant jet engine developments during that time. The original USAF requirement was for a strategic nuclear bomber that did not depend on foreign-controlled air bases. That meant it had to have a long range. The first design, the 1946 Boeing Model 462 prototype, featured Wright T35 turboprops on straight wings with huge propellers. The USAF realised this would be obsolete before it took off and asked for something more radical. The result was a new design using the swept wing of the B-47 and the eight Pratt & Whitney JT57 twin-spool jet engines.

Fuel consumption was still a concern, and in 1961 the B-52 H, the last variant built, had the new Pratt & Whitney TF33-P-3 turbofan engines installed. These had reliability problems but provided much better performance and fuel economy than the J57 turbojets.

New Rolls-Royce F 130 engines will be fitted to the 76-strong fleet soon, still eight per ship, improving efficiency by another 20 per cent. These will power the Stratosaurus for another 30 years.

As we saw earlier, turbofans were another major advance in jet engine technology, but they had been thought of years before by the engine's inventor. Whittle had realised that it was preferable to obtain thrust from a jet engine by having a large airflow rate and a relatively small jet velocity.

In the first jet engines, the exhaust was far faster and hotter than it needed to be for the most efficient thrust. That is partly why the first Comet airliners were so fuel-thirsty and relatively lacking in power. You might compare it to a tiny propeller on a boat with an outboard motor that has to revolve frantically fast to have the same thrust as a larger propeller revolving lazily. Capturing some of that wasted energy would improve the efficiency of the engine, resulting in a longer range, reduced fuel bills and increased power.

The Wright T35 turboprop engines originally proposed for the B-52 used a series of turbine stages to harvest all the exhaust energy and direct it to a propeller. This is why turboprops are more fuel-efficient and have a better low-speed take-off performance. As we have seen, though, propellers decline in efficiency as they approach 500mph (800km/h, 430 knots). There was, therefore, a sweet spot to be exploited between the turboprop at low speeds and the pure jet at high speeds. This sweet spot, between 450mph (720km/h, 390 knots) and 700mph (1,100km/h, 610knots), is exactly where jet passenger aircraft spend so much of their working lives. So the potential gains of a hybrid between the two were enormous.

It is clear that Whittle had foreseen all this in his very first patent application. In January 1930, he had drafted his Patent Provisional Specification No. 347,206: 'Improvements relating to the propulsion of aircraft and other vehicles.' As well as proposing the new jet technology, he had almost casually referred to two further developments in the field. One was the concept of a bypass design. Most of the air from the compressor is made to bypass the combustion chambers and the turbine and exit the back of the engine where it surrounds the high-velocity jet exhaust. Because this air hasn't been accelerated by combustion, it exits at a lower speed than the jet exhaust, a speed closer to the aircraft's airspeed, therefore improving the Froude efficiency. This is defined as the ratio of useful power output to the rate of energy input. There is a further advantage. Residents around airports had become infuriated by the noise of the first jet engines and the bypass design surrounded the noisy hot core of jet exhaust with a benign, calming sleeve of slower moving cold air, which reduced the noise, somewhat.

Furthermore, in the same patent application, Whittle suggested that jet engines could be 'directionally controlled' to provide vertical take-off capability. This idea would eventually bear fruit in the form of the 1967 Hawker-Siddeley Harrier. It's remarkable that this 22-year-old genius had foreseen these two later developments in jet propulsion technology.

In 1946, Griffith, who was by then the chief engineer at Rolls-Royce, proposed the construction of a bypass engine based on the Avon. And so Rolls-Royce combined axial-flow, twin-spool and bypass ideas and built its Conway engine, named after a river in North Wales.

Air molecules loitering around at high altitude would first become aware of the Conway as they were sucked into the first of seven stages of a low-pressure compressor and squeezed tighter and tighter. The first six stages of blades were made of aluminium and the last was made of titanium.

Next, the air was compressed even further by the high-pressure compressor, spinning at a higher speed due to the twin-spool design. This had nine stages, the first seven made of titanium and the last two stages made of steel. By now, the molecules had been compressed 14 times. At this point, 30 per cent of the air was directed into bypass ducts, also made of titanium, and led round the hot section of the engine. The remaining 70 per cent was pumped into ten cannular combustion chambers, or flame cans made of Nimonic alloy. Here the air was mixed with kerosene and burned. The single-stage turbine blades, which were also made of Nimonic alloy, were hollow and air-cooled and drove the high-pressure compressor. After that, a two-stage turbine drove the low-pressure compressor. Then the hot exhaust gases jetted out the back of the engine to be reunited with the cooler, slower bypass air. The air molecules could then return to their high-altitude existence, somewhat shaken – and somewhat hotter.

The Rolls-Royce Conway developed 17,500lbf (78,000N) and had a specific fuel consumption of 0.725 at take-off and 0.874 during the cruise. These were excellent figures.

To make the engine compact, the accessories were mounted at the front, and the oil was contained in a wet sump. This engine was fitted to the Boeing 707, Douglas DC-8-40, the Handley Page Victor Cold War bomber and the rear-engined Vickers VC10. This last airliner's Conway engines had carbon-fibre fan blades fitted in 1968, an innovation that would soon cause headaches at Rolls-Royce.

The VC-10 was designed to cope with the 'hot and high' African airports that were of such concern to the Brabazon committee. To deal with the shorter, rougher runways, it mounted its four powerful Rolls-Royce Conways high up in the tail, away from debris, and had large low-pressure tyres to deal with the bumps. It was a successful airliner, popular with passengers and crews. It was fast, too, performing the fastest crossing of the Atlantic by a subsonic jet airliner of 5 hours and 1 minute, a record that was held for 41 years. Only Concorde was faster. Unfortunately, BOAC's preference for buying American planes never gave the VC-10 a chance.[1]

One VC-10 was borrowed by Rolls-Royce to test its RB-211 engine. When returned, it was found that the airframe was distorted,

---

1 In 1956, BOAC ordered 15 Boeing 707s, and then 11 Boeing 747-100s in the late 1960s. BOAC was then known as 'Buys Only American Companies'.

possibly due to the power difference between the RB211 on one side and the Conways on the other. They had bent it!

In 1961, Pratt & Whitney, hearing of the Rolls-Royce findings, swiftly developed the JT3D turbofan. Over 8,000 JT3Ds were eventually produced between 1959 and 1985. They were so superior in performance that replacements of the old JT3C on Boeing 707s and DC-8s were made as fast as Pratt & Whitney could turn them out. The results were staggering: take-off thrust increased from 12,500lb from the old engines to 17,000lb from the new. Not only that, but they also used 13 per cent less fuel.

This meant that, when fitted to the B-52, the aircraft could carry more bombs even further.

Among aircrew, the B-52 is known as the BUFF: the Big Ugly Fat F***er. What's it like to fly? 'Like flying a museum,' says USAF pilot Carlos Espino. 'It's a brick: I would say it's like wrestling.' The two pilots sit side-by-side in a tiny cockpit surrounded by analogue gauges. 'It has a lot of redundant systems,' Espino adds. 'So if one system fails, there's plenty of other systems to back it up.' For example, the eight engines mean that if one quits, there isn't a huge asymmetrical imbalance as there would be with a twin-engined aircraft. The hardest manoeuvre, he says, is lining the huge bomber up with a flying air tanker: 'At the end of air refuelling, you're literally sweating.'

In Stanley Kubrick's *Dr. Strangelove or: How I Learned to Stop Worrying and Love the Bomb*, a rogue B-52 drops a hydrogen bomb on a Soviet missile base. One of the sets was a recreation of the B-52's cockpit but, as the bomber was on the secret list, the Pentagon was not keen on letting the makers of a comedy film loose in its aircraft. The set designers had one snatched photograph to go on, but when they invited USAF staff to the set, they were told they had got it spot on, even down to a small box that was the highly secret recall device. 'It was absolutely correct, even to the little black box which was the CRM,' reported one anonymous source. It was this box that failed to call off the last B-52 from the raid in the film. On the bombing run, the crew are unable to open the bomb-bay doors, so the commander, Major 'King' Kong, climbs into the bay, straddles the bomb and fixes the doors. They open, and he rides the hydrogen bomb, rodeo-style, down to destruction. *Dr. Strangelove* was voted 26th in a list of the best-ever American movies.[2]

The B-52 takes off with huge plumes of black smoke from the eight exhausts. Behind the pilots is a tiny compartment with a navigator and radar operator. Unlike the British V-Force bombers, these aircrew have ejector seats. These fire downwards, though, so safe evacuation height is above 250ft.

Once aloft, the Stratosaurus can just keep going. A B-52H once set a world distance record by flying unrefuelled from Japan to Spain, covering 12,532 miles (20,168km).[3] That's halfway around the world. During the Cold War, airborne B-52s used to loiter near the borders of the Soviet Union, loaded with nuclear weapons and ready to dash to their targets, either for a rapid first strike or retaliation.

But there are problems with flying around with live nuclear weapons, and the Boeing B-52 Stratofortress also suffered broken arrow incidents. In 1961, a B-52 carrying two 3-megaton Mark 39 nuclear bombs broke up in mid-air, dropping them near Goldsboro,

2 The American Film Institute (1998)

3 10,895nm.

North Carolina. One bomb came close to detonating: three of the four arming mechanisms activated. Just one cheap electrical switch out of the four prevented a major nuclear catastrophe, one which would have changed the face of the United States.

In 1966, a KC-135 tanker and a B-52 collided during refuelling. The tanker's fuel load blew up and the B-52 disintegrated, dropping four Mark 28 nuclear bombs onto the fishing village of Palomares in Almería, Spain. One fell into the sea and was eventually recovered after a ten-week hunt.

And then, in January 1968, a crewman in a B-52 over Greenland stuffed seat cushions over a heating vent, causing a fire. The crew had to eject, leaving the aircraft and its load of four hydrogen bombs to make its own landing. The wreckage spread over the sea ice of Wolstenholme Fjord in the northwest corner of Greenland. The US wanted to let the bombs sink through the ice, but Denmark insisted that the wreckage was all retrieved, which it was successfully except for one cylinder of uranium and lithium deuteride. Which is still there, somewhere.

What is astounding is that there haven't been more major accidents involving nuclear weapons. The Soviet Union probably had as least as many broken arrow incidents but chooses not to make them public.

One of the more disturbing developments involving jet engines and atomic technology was the nuclear-powered aircraft developed by both sides during the Cold War. A nuclear reactor heats compressed air in a turbojet using fission, not fossil fuels. An aircraft thus equipped can fly indefinitely with no refuelling, always ready to attack. The Americans built the experimental Convair NB-36H bomber, which mounted no less than three types of power plant: six Pratt & Whitney R-4360-53 radials, four General Electric J47 turbojets and one 1-megawatt nuclear reactor in the bomb bay. This aircraft was used to measure the amounts of radioactivity emitted by the reactor during flight, in an experiment called Project Halitosis. The NB-36H remains the only American aircraft to have carried an operational nuclear reactor.

Later, the Americans built a nuclear ramjet test engine called Tory-IIA and attached it to a railway carriage. It was fired up on 14 May 1961, probably becoming the world's first nuclear-powered jet-engined railway locomotive. And the last, I hope.

One of the many problems of flying around with an open nuclear reactor is the heavy shielding. The Convair's weighed 11 tons. This was solved on the Russian Tupolev Tu-119 by simply not having any shielding. It had two turboprop engines and two direct-cycle nuclear jet engines. The test pilot, E. A. Guryenov, reported: 'We had all been irradiated, but we ignored it. Of the two crews, only three men survived: a young navigator, a military navigator and me. The first to go, a young technician, took only three years to die.'[4]

It is hard to uninvent dangerous technology, and once invented it tends to be used by the desperate or the unscrupulous. There is also the principle of Chekhov's gun. The Russian playwright stated, 'If in the first act you have hung a pistol on the wall, then in the following one it should be fired. Otherwise, don't put it there'[5] He meant that writers should remove everything irrelevant to the story, but it also serves to remind us that once a weapon is produced, it will be used.

The nightmare prospect of atomic reactors flying above our heads, trailing a plume of radioactive particles, has not receded. In 2018, Russian president Vladimir Putin proclaimed that Russia had

---

4 In a letter to journalist George Kerevan.

5 Recorded by Ilia Gurlyand in *Reminiscences of A. P. Chekhov* (1904). Hemingway, with his love of inconsequential detail, disagreed and introduced two characters in his short story *Fifty Grand* who are never again mentioned.

developed a nuclear-powered cruise missile with a nuclear warhead that could evade air and missile defences and hit any point on the globe. The 9M730 Burevestnik has, he said, a virtually unlimited range.

Then, on 9 August 2019, a Russian nuclear energy agency admitted that there had been an accident killing five scientists. This may have involved a failed test of the Burevestnik missile, which remains at the bottom of the White Sea. On 21 November 2019, Putin said the scientists had been testing an 'unparalleled' weapon: 'We are talking about the most advanced and unparalleled technical ideas and solutions about weapons design to ensure Russia's sovereignty and security for decades to come... [the] weapon is to be perfected regardless of anything.'[6] In the light of his invasion of Ukraine, these are chilling words indeed.

The US owns 70 B-52s and they're still on the front line. 'The B-52 has been a workhorse of the Air Force for decades,' said Todd Harrison of the Center for Strategic and International Studies.[7] 'It's a remarkable aircraft, and I think it has really proven out the concept that your major platforms can stay relevant, long after their design life, by upgrading the components and the technologies that go on them.'

The next big upgrade is yet another change of jet engines, which will cost around US$130 million per plane, and it will mean that the 100-year-old airframes will have seen four generations of jet technology. This will yield another 20 per cent gain in efficiency. But it is still old tech: 'If we have a conventional fight against Russia

---

6 https://www.themoscowtimes.com/2019/11/22/putin-says-unparalleled-weapons-tested-at-deadly-nuclear-accident-site-a68274

7 https://www.popsci.com/story/technology/inside-air-force-b52-training-mission

or China,' said Harrison, 'the B-52 is a sitting duck to air defence systems and to Chinese and Russian fighter jets.' The BUFF wouldn't survive more than a few minutes over a modern well-defended state.

The Boeing B-52H Stratosaurus bomber was used mostly for carpet-bombing the unfortunate citizens of developing countries. When the United States started its secret war with Laos during the Vietnam War in the 1960s and '70s, the people on the ground didn't stand a chance. The 288 million cluster bombs that were dropped are still killing and maiming Laotian children, who find the tennis-ball-sized bombs irresistible. The bombs 'fell like rain', US President Barack Obama was told on his visit to Laos in 2016.

'Given our history here, I believe that the United States has a moral obligation to help Laos heal,' the president said.

He described Laos as the most heavily bombed nation in history. Eight bombs a minute were dropped on average during the Vietnam War between 1964 and 1973, more than the quantity used during the whole of the Second World War. The US flew 580,344 bombing sorties over Laos, dropping two million tons of ordnance, with many targets in the south and north struck repeatedly as part of efforts to isolate Communist North Vietnamese forces. As they neared the border, any remaining bombs were simply dumped.

Most devices dropped were anti-personnel cluster bombs, and an estimated 30 per cent of these munitions did not detonate. Around 29,000 people were killed and 21,000 injured. The overwhelming majority were civilians, among them thousands of children who are still dying 50 years later.

Chapter 15

Sangri-la

During the Second World War, many Allied service personnel ended up seeing rather more of the world than they bargained for, but if they survived they returned home with an increased appetite for foreign travel. Preferably without being shot at.

Jet travel was still beyond the means of ordinary people, and newspaper reports of the antics of the Jet Set showed them what they were missing. Jet airliners were expensive and national airlines were determined to keep the monopoly to themselves and their ticket prices high. The Brabazon committee had indicated that air travel was for government officials and for the wealthy, not the ordinary citizen. The post-war Labour government enforced this by imposing tight exchange controls, effectively banning leisure travel abroad for three years after the war. British European Airways (BEA) was owned by the government and was determined to keep their monopoly. But the public's appetite to travel was still unsatisfied.

Enter Vladimir Raitz, the inventor of the package holiday.

In his excellent account of the low-cost revolution in the skies, *No Frills*,[1] travel journalist Simon Calder tells us how it all happened. Raitz was born to a White Russian[2] Jewish family in

1 Calder, Simon, *No Frills*, Virgin Books, London (2006)

2 White Russians fought against the Soviet Red Army during the Russian Revolution (1918 to 1921).

Moscow. They fled Russia in 1928 and ended up in London, where the young Vladimir attended the London School of Economics (LSE) – the same college as Stelios of easyJet, Mick Jagger and Cherie Booth.

Then, in 1949, aged 27, he was invited to a holiday in Calvi, Corsica. 'The property was called Club Olympique,' remembered Raitz, 'and it was run by White Russian friends of mine.' Here he was asked by a local socialite, Nicholas Steinheid, if he could encourage more British visitors to come the following year.

Using his LSE business skills, Raitz calculated that he could charter an aircraft and give the tourists an all-inclusive two-week holiday in Corsica for under £35. This would avoid the British government exchange controls, as his customers wouldn't have to carry much cash. This was still a lot of money, about £1,225 today, but it was still less than half the cost of a return flight to Nice by BEA. It wouldn't be luxurious, as his clients would have to sleep under canvas in ex-US Army tents left over from the war. So, financed by £3,000 left to him by his grandmother, Raitz set up Horizon Holidays in October 1949 – and kicked off the package holiday industry.

As he expected, the British government was obstructive. His civil aviation licence didn't come through until March 1950, and when it did it limited passengers to 'students and teachers only', as travel to Corsica was deemed to be only of educational value. Other occupations were banned. There were also strict government regulations about how much money you could take on an overseas holiday: 'At first it was £50, and that was later cut to £25,' Raitz remembered. 'But we could keep the price of accommodation down to about £15, and the charter flight was payable in sterling, so there was still some spending money left over.'

Raitz ran an advertising campaign in teaching and nursing magazines offering a flight, plus 'delicious meat-filled meals and as much local wine as you can put away' – quite an incentive in austerity Britain in the 1950s. The final cost of the first package was £32 10s. Horizon's first passengers took off from Gatwick for the sun in May 1950.

The aircraft wasn't jet-engined; it was our old friend, the 32-seater propeller-driven Douglas DC-3 Dakota, a government-surplus aeroplane. The charter of the aircraft had cost Raitz £305 for the return flight: about £10 per passenger. And the Dakota was slow; at 170mph (270km/h) it was even slower than a modern high-speed train. It had to stop at Lyon for fuel, but Raitz turned the obligatory stop into an attraction, offering his clients a meal in the gastronomic heart of Europe. For many of his passengers, this would be their first taste of a foreign country. Then, after an eight-hour journey, they arrived at Calvi. Nowadays that flight would take just two hours.

'When we arrived at Corsica airport, there was nothing at all,' said Raitz 'not even a little hut. We had to shelter from the sun under the wings of the plane while we waited for the bus to pick us up.'

They were then bussed to the camp at Club Franco-Britannique. Once at the beach, the holidaymakers found large canvas tents, each with two beds. There was one area set aside for 'ablutions', a dining room, bar and dance floor. 'A pastis was a few pence,' Raitz said, 'and a bottle of wine was nine pence.' The locals were welcoming, turning out the town band and garlanding the tourists with flowers.

The first season was popular with his clients, but Horizon Holidays didn't make much money. The next season was better, as the incoming Conservative government removed some of the restrictions. Even so, Raitz had to fight hard for every licence: BEA opposed him at every hearing. They objected to his flying to Corsica because it constituted 'material diversion of traffic', even though they did not fly to the island!

One of the rules was particularly restrictive; on routes that BEA or any other scheduled airline flew, the minimum price Raitz was allowed to charge for a whole package holiday was the lowest fare set by the airlines' cartel, the International Air Transport Association (IATA). So, in the above example of the BEA return ticket to Nice for the modern equivalent of £2,500, Raitz would have to charge at least that sum for a package holiday. Which would not get him many sales.

This would be considered an outrage today, but back then there were no consumer champions like Simon Calder to fight their corner for them.

Vladimir Raitz persevered, expanding his operation to the Costa Brava, then to Mallorca: 'When I went there, there was only one hotel on the island. There were no roads, you had to clamber down a mountainside to see a lovely little beach. That was what was so satisfying.' Sardinia followed, then Malaga and Perpignan.

In those innocent days, little thought was given to the blight of modern mass tourism, nor to the effect on local populations. In Spain, the youngest sons who had inherited the worthless lands along the seashore found themselves hotel-owning millionaires, while the elder brothers continued to scrape a living on the family plots inland.

Once Vladimir Raitz had shown the way, dozens of other companies sprang up, jet aircraft were chartered instead of old piston-engined Dakotas and the accommodation improved. Raitz was proud of what he had achieved:

> Providing a fortnight in the Mediterranean sun to a wide segment of the British public, hereto the prerogative of well-to-do members of the bourgeoisie, brought with it what can only be described as a social revolution; the man in the street acquired a taste for wine, for foreign food, started to learn French, Spanish or Italian, made friends in the foreign lands he had visited; in fact became more cosmopolitan, with all that that entailed.

What about the destructive effects of tourism? 'Benidorm' he said in 1989, 'looks bloody awful now – but that's progress, I suppose…

On one hand, I hate to see resorts being despoiled. Take Minorca. There used to be a beach with no road to it… today it's ringed with hotels. To that extent, I am sad… On the other hand, I think it's marvellous that 12 or 13 million people can have a Mediterranean holiday and enjoy themselves.'

There was a difference in clientele: 'In the early days, we were a middle-class, quiet crowd. No lager louts. And, as far as my companies were concerned, we never did have any – even though I invented Club 18-30 too, in 1970.'

Just before Horizon's third summer season, on 1 May 1952, TWA introduced a 'tourist' class. Before that, all seats on scheduled airlines were effectively first class, with ticket prices to match. TWA had begun to see the possibilities of piling high and selling cheap. And, as we have seen, the following day a de Havilland Comet took off on the first-ever passenger jet service, from London to Rome.

The stage was now set for an explosion in cheap jet travel. By 1960, companies were able to offer two weeks in Majorca for £41. A fortnight in St Remo was £46, a ten-day tour of the Swiss Alps £54 and a grand tour of Rome, Venice, Florence and Capri £68.

With the exponential rise in tourist numbers came tumbling prices: a package that would have cost £80 in 1951 was the same price despite inflation in 1971. And between 1969 and 1993, the cost of a fortnight in Torremolinos halved in real terms.

By 1960, 3.5 million Britons took overseas holidays, by 1979 it was 10 million and by 1987 it was 20 million. In 2019, nine out of ten UK residents took 72.6 million trips overseas in total.

In fact, according to data from IATA, Britons are the most prolific international jet travellers. In 2018, 126.2 million passengers were British – totalling 8.6 per cent, roughly one in 12 of all international travellers. However, in 1972, Raitz left Horizon after it was taken over by Court Line, which in turn went into liquidation in August 1974.

If Vladimir Raitz made jet holidays affordable, then Freddie Laker made jet flights cheap. And the key to that was powerful and reliable jet engines. As we have seen, in the US the commercial aviation network had been set up for the transport of US mail. As a result, the nation had been carved up between a few big airlines, and there was little competition between them for around 40 years. And they were perfectly happy about that. The federal government didn't encourage competition across the nation as it was regarded as a duplication of service and thus a waste of resources. Competition was therefore restricted within states, with only three big enough to be worth operating airlines within: Alaska, California and Texas. Alaska had only one resident per square mile, so if competition between airlines was going to start, it would start in California or Texas.

Pacific Southwest Airlines (PSA) started up in California in 1949 and offered flights from San Diego in the far south to Burbank, Los Angeles and Oakland for San Francisco. They learned that the way to make money was to undercut the established competition (not difficult) and to keep their aircraft in the air as much as possible. When they bought jet-engined planes, they found they were more reliable and made even more money. Southwest Airlines followed PSA's model and did the same in their home state of Texas, flying between Houston, Dallas and San Antonio. They had to fight the established airlines tooth and nail in the courts even to get started. Staying within their home states, both cut-price airlines prospered.

Freddie Laker, an English entrepreneur, decided in 1971 to take on the big national airlines crossing the Atlantic by starting a low-cost no-frills airline he called Skytrain. Seven years of legal battles on both sides of the ocean followed. In 1973, Skytrain was designated in the UK as a transatlantic carrier. More legal problems emerged in the US, where Pan Am and TWA fiercely resisted Skytrain's proposed competition. Eventually, in September 1977, Skytrain launched the world's first daily transatlantic low-fare scheduled service between London and New York City, charging just £32.50 in winter and £37.50 in summer. This was just one-third of what the established 'flag carriers' were charging at the time.

Freddie Laker was knighted the following year, and his model served for Virgin Atlantic and easyJet and all the low-cost airlines that followed.

I, for one, am grateful for the affordable travel he opened up in my life, and if anyone deserved a knighthood, it was Laker. 'Freddie Laker was a man who had courage and vision,' said one US airline executive. 'He was brave.'

On 26 September 1977, the Laker Skytrain launched between Gatwick and New York JFK, rapidly expanding to Miami and Los Angeles. A young Richard Branson joined the 'mile-high club': 'I was sitting in economy on a Freddie Laker flight, next to this very attractive lady, as we headed to Los Angeles,' boasted the future boss of Virgin Atlantic. 'We got chatting and it went a bit further.'

Sadly, Laker Airways went bankrupt in 1982, owing over £250 million (£887 million today). It was the biggest corporate failure in Britain up to that time. Laker Airways had expanded too quickly when it took on a fleet of Douglas DC-10s. When these aircraft were grounded as a result of an American Airlines crash at Chicago, it lost £13 million. Building a business on discount travellers was difficult, and there was a concerted campaign from the established airlines to price match Laker Airways, even at the cost of huge losses. In a later interview in 2001,[3] Simon Calder asked Sir Freddie, 'Why did they do that?' Freddie replied, 'The real reason they went for me

---

3 Calder, *No Frills*

at the jugular was because of our application for Europe. With the European Union and everything else, we were going to get freedom of the skies in Europe... imagine having Laker in Europe 22 years ago. How much is that worth? Billions. Billions.' Sir Freddie was still bitter, and rightly so. 'It's all documented out there, what the bastards tried to do.'

In the end, Sir Freddie, like Sir Frank Whittle, like Vladimir Raitz, suffered from first mover's disadvantage. Those coming after the pioneers benefitted, but pioneers of all sorts should be celebrated for the heroes they were. Without the jet engine, though, they could never have brought us cheap mass transport.

Chapter 16

Skyjack

The continually increasing power and reliability of the jet engine led to the realisation that four engines were no longer needed for some routes. Deleting one power plant would save airlines millions of dollars and drive ticket prices further downwards. One reason for the demand for tri-jets was the Extended-range Twin-engine Operational Performance Standards (ETOPS) regulations. This was called the '60-minute rule' by the Federal Aviation Administration (FAA).

With the sorry history of unreliable piston-engined planes behind them, these rules stated that twin-engined aeroplanes were only allowed to fly on routes that were within one hour from the nearest airport. If one engine failed (as so often they did), there would be enough time to make an emergency landing on the one remaining engine. Long-distance routes over the oceans were, therefore, not possible for twin-engined aircraft.

With the rules allowing three engines, tri-jets became possible on intercontinental routes and were chosen by airlines as being not only cheaper to buy but, without the drag of four podded engines, also more fuel-efficient. The three-engined wide-bodied jets were seen as a good compromise between twin and quad-jet aircraft, having better range, payload and passenger capacity than twin-jets without the huge fuel consumption of quad-jets. Later changes in ETOPS rules would see the allowance rise from 60 minutes to 120 minutes, and then up to 180 minutes. ETOPS-180 was enough for manufacturers to shift their development towards today's long-range twin-jets.

In July 1956, BEA had announced a request for a short-haul 'second-generation jet airliner'. Four British companies started a furious competition: Bristol, Avro, Vickers and de Havilland. The latter came up with a design featuring a high T-tail and three rear engines. Pure jet engines, unlike most propeller-driving engines, are slim enough to fit a centre-mounted jet engine in the tail. It necessitated an S-shaped duct to provide the centre engine with air, and this caused flow distortions which, in turn, made the centre engine surge. This was solved by building in vortex generators in the first bend of the duct.

Dithering by the British government and its forcible consolidation of the British manufacturers delayed the appearance of a prototype. But the first three-engined jet airliner to be designed was the de Havilland version, soon to be named the Hawker-Siddeley HS-121 Trident.

Meanwhile, Boeing had been watching the competition and started its own tri-jet programme in 1959. Lord Douglas of BEA hoped Boeing might drop its plans and build the Trident in the USA. With astonishing naivety, the British invited a team of Boeing engineers to Hatfield and showed them all their plans and research.

'De Havilland solemnly handed all its research over to its rivals…' commented aviation author Derek Wood, 'the crowning piece of stupidity.'[1]

In the end, only 117 Tridents were produced, although it was a successful design. And no less than 1832 Boeing 727 aircraft (designed to the Trident's original specification) were built.

---

1 Wood, Derek, *Project Cancelled*, Macdonald and Jane's Publishers, London (1975)

The Boeing designers knew that their aircraft had to meet American demands, with six-abreast seating and the ability to use short runways as little as 4,500ft (1,370m) long. This would enable the new 727 to use the small airports scattered all across the US.

The 727 was made for these airports, and independence from ground facilities was a vital requirement. This was also the reason behind the innovative rear door and stairway that hinged up from the belly of the aircraft beneath the high-mounted centre engine.

This built-in airstair could be opened in flight, and that led to the extraordinary story of D. B. Cooper, the mysterious hijacker who used this exit when he parachuted from the tail of a 727 as it was flying over the Pacific Northwest.

The first years of post-war jet travel were a delight. Airport security was negligible and a duty-free shop was a place where you would actually want to buy things. However, after the Cuban Revolution, there was an explosion of hijackings by passengers who wanted to fly to Utopia. As a result, in 1968, more than 1,000 Americans had an unplanned trip to Cuba. In 1969, there were 86 skyjackings and, in the five years between 1968 and 1972, more than 130 aircraft were hijacked in the US alone, sometimes two on the same day.

For the other passengers, it meant landing in Havana and a long delay. For the pilots, it meant carrying charts of the Caribbean Sea and Spanish phrase cards reading: 'I must open my flight bag for maps' and 'Aircraft has mechanical problem. We can't make Cuba.' For the airlines, it meant expense, because Fidel Castro was charging US$7,500 a time to retrieve their planes. But what was in it for the hijacker?

'In a few hours it would be dawn in a new world… I was about to enter Paradise,' one skyjacker remembered thinking. 'Cuba was creating a true democracy, a place where everyone was equal, where violence against blacks, injustice, and racism were things of the past. I had come to Cuba to feel freedom at least once.'

The hijackers saw themselves as part of the struggle: 'Most skyjackers earnestly believed that upon reaching Havana, their sole destination during the mid-to-late Sixties, they would be greeted as revolutionary heroes,' Brendan Koerner wrote.[2] 'Every skyjacker was an optimist at heart, supremely confident that his story would be the one to touch Castro's heart.'

However, Fidel Castro had nothing but contempt for these people, whom he regarded as troublemakers. After landing at Havana's José Martí airport, hijackers were dragged to the headquarters of G2, the Cuban secret police. Here they were accused of working for the CIA and interrogated for weeks on end. If they angered the interrogators, they were sent to work in punitive sugar-harvesting camps. Once there, they were regularly flogged, and if they complained, they were executed. If they escaped, they were recaptured and dragged across the sharp stalks of sugarcane until the flesh was ripped off their bones.

The lucky ones found themselves in Hijackers House, or Casa de Transitos, where up to 60 inmates were crammed into small rooms and forced to live on 40 pesos a month.

Despite rumours of this brutality filtering back to the US, the hijackings to Cuba continued. One involved a man dressed as a cowboy, who turned out to be the 28-year-old heir to a real-estate fortune. Another wanted to study communism at first hand, and yet another hijacker wanted to return to the island because he missed his mother's frijoles. All three were probably disappointed.

The US government grew exasperated by the hijackings and canvassed public opinion. Trapdoors in the cockpit, perhaps, with a lever so the hijacker could be deplaned in flight? How about this

---

2 Koerner, Brendan, *The Skies Belong to Us: Love and Terror in the Golden Age of Hijacking*, Crown, New York (2013)

idea: all passengers had to wear boxing gloves so they couldn't hold a gun? Another idea was to play the Cuban national anthem before every flight, with anyone knowing the words being arrested. How about free one-way flights to Cuba? Castro was not so keen on this last idea, which he dismissed as 'good riddance flights'. The favourite suggestion was to build a perfect copy of Havana's José Martí airport in Southern Florida and to land planes there.

The FAA was eventually given the job of solving the problem. At a senate hearing in 1968, the FAA's Irving Ripp said hijackings were an 'impossible problem, short of searching every passenger'. 'If you've got a man aboard that wants to go to Havana, and he has got a gun, that's all he needs,' he said.

This defeatist notion, surely only possible in gun-carrying America, was roundly dismissed by Senator George Smathers of Florida, who suggested using metal detectors or X-ray machines to screen passengers, just as used at prisons for visitors. 'I see no reason why similar devices couldn't be installed at airport check-in gates to determine whether passengers are carrying guns or other weapons just prior to emplaning,' Smathers insisted.

Irving Ripp wasn't so sure. X-ray screening would have 'a bad psychological effect on passengers... It would scare the pants off people. Plus, people would complain about invasion of privacy.'

Hijackers grew bolder. 'Through the end of '71 and '72, hijackers were demanding hundreds of thousands of dollars, or sometimes gold bars – lots of material wealth in exchange for the passengers,' remembered Koerner.

Finally, on 10 November 1972, there was a hijacking that made the US government wake up to the huge potential dangers. Three criminals hijacked Southern Airlines flight 49 and threatened to fly it into the nuclear reactor at the Oak Ridge National Laboratory in Tennessee unless they received a multi-million-dollar ransom. This they collected, then ordered the pilot to fly to Havana. Here though, they were arrested at gun point and served eight years in a Cuban prison before returning to the US to serve additional 20-year prison sentences.

Within two months, the FAA had introduced passenger screening, bag searches and metal detectors. That was the beginning of airport security as we know it. Hijackings are rare now, with none in 2015 and 2017.

There have, though, been cases of hijackers in Cuba taking control of planes on the island and demanding to be flown to the US.

The man sitting in the back of the Boeing 727 slipped a note to one of the flight attendants, Florence Schaffner, shortly after take-off from Portland airport. Assuming it was a note declaring love, she put it into her purse without reading it and walked on. When she returned, the man flagged her down and said 'Miss, you'd better look at that note. I have a bomb.'

He motioned her to sit down beside him and opened up his briefcase. Inside, Schaffner saw eight red cylinders, four on top of four. He closed the case, took back the note and stated his demands: he wanted $200,000 'in negotiable American currency', four parachutes (two primaries, two reserves) and a refuelling truck waiting for him upon landing. The man, who had given his name at the ticket counter as Dan Cooper, asked for the money to be in $20 bills.

It was 1971, before the introduction of passenger screening. Cooper had just carried his bomb-briefcase straight on board. Standing instructions to aircrew were to accede to hijacker's demands, so Flight 305 circled Seattle until the parachutes and money were collected. Schaffner remembered that Cooper was familiar with the terrain they were overflying: he mentioned that an air force base was a short drive away from the airport they were flying to, and he recognised the Tacoma River when they passed over it.

Cooper allowed the other passengers to disembark. While the 727 was being refuelled, he briefed the crew. They were to fly south towards Mexico City at the aircraft's minimum speed, just above stall: about 100 knots. The altitude was to be 10,000ft, the undercarriage was to be lowered, the wing flaps lowered to 15° and the cabin unpressurised. The crew protested that these conditions would lead to heavy fuel consumption and that they would have to land at Reno, Nevada, to refuel. Cooper agreed to this. The money (which had secretly been photographed) and the parachutes were loaded on board.

After take-off, Cooper ordered the crew and flight attendants to stay in the cockpit with the door closed. Then, after night had fallen, around 2000hrs, in mid-flight a warning light appeared in the cockpit: the 727's rear door had just opened. This was accompanied by a noticeable change in air pressure. Then, a few minutes later the pilot felt the aircraft's tail lurch upwards, and he had to re-trim the 727 back to level flight.

Cooper had taken his big jump into the darkness.

The mystery obsessed America. None of the five aircraft tailing Flight 305 saw the black-suited Cooper jump. After landing at Reno, armed police searched the 727 but few clues were ever found.

The most intensive search and recovery operation ever undertaken in the US was mounted, but nothing significant was found until nine years later, when a child found some bundles of the ransom cash on a bank of the Colombia River, Washington.

It was known that during the Vietnam War the CIA had rehearsed clandestine parachuting from the Boeing 727's back door. The FBI, with penetrating deduction, drew a profile of Cooper as a former military man who lived in the Northwest, and who was a capable outdoorsman. Even with this inside information, Cooper has never been found.

This remains the only unsolved hijacking on US territory, and what none of the articles, books and films about the mystery point out is that it was the novel positioning of a jet engine that enabled Cooper to jump out of the back of a Boeing 727.

Then, on the morning of 11 September 2001, 19 hijackers took control of four Boeing jet airliners, and the world changed again.

The date chosen may have been significant: during 1683, Muslim armies were attempting to capture Vienna, the seat of the Holy Roman Empire, and on 11 September of that year the battle began, which ended with the West gaining dominance over Islam. By attacking on this date, Osama bin Laden perhaps hoped to symbolise the beginning of Islam's resurgence.

Large aircraft with long flights were chosen for the 9/11 attacks because their tanks would be full, and it was the combustion of hundreds of gallons of jet fuel that brought down the World Trade Center towers, not the initial impact of the Boeing 767s. The jet engines that produced all the horror and the carnage were General Electric CF6 high-bypass jets.

In total, 2,996 people died in the 9/11 attacks. To put this into context, around 3,100 people die on American roads every month, and about the same number were dying every day at the height of the Covid pandemic. The difference, of course, was that this was mass murder, and it inflicted a huge psychic wound on the American nation.

In the aftermath of the 9/11 attacks, North American air space was closed for several days, and air travel declined after reopening, leading to a 20 per cent cutback in air travel capacity. The plane-free skies were an opportunity to study the climate warming effect of the condensation trails that form in the wake of high-flying jet engines. It is still unclear whether these contrails increase the total amount of clouds in the sky or suck up moisture that might have enabled clouds to form elsewhere. As with much climate science, it's complicated.

The Swedish environmentalist Greta Thunberg famously refuses to fly, choosing instead to sail across the Atlantic to New York in a yacht. But this vessel was built out of unsustainable petro-chemicals and was crewed by sailors who took long-distance jet flights on their return.

The very ease and affordability of jet air travel has produced the agenda that Thunberg promotes: that people across the world must unite to combat climate change. But that consensus would never have come about if it were not for people being able to fly to other countries, see other cultures and feel a sense of being part of a global community instead of a narrow national identity. Thunberg owes the power of her message to the very technology she berates. As I said, it's complicated.

Chapter 17

Concorde

'None of the other nations of Europe has so abject an inferiority complex about its own aesthetic capabilities as England,'[1] wrote the German-born British art historian Nikolaus Pevsner in 1956. And yet England has contributed much, not only the country houses that Pevsner so admired but also some of the most beautiful aircraft ever built: Spitfire, Comet… and Concorde.

We have already met Johanna Weber, the German-born mathematician and aerodynamicist who, together with fellow German Dietrich Küchemann, worked on the Handley Page Victor V-bomber. The two aerodynamicists had moved to England in 1947 after the end of the war. The airflow was so efficient over the Victor's crescent-shaped wings that their design was also used for the Airbus A310 introduced in 1983.

In 1956, Weber and Küchemann started to work for the government-funded Supersonic Transport Advisory Committee (STAC) at the RAE. Weber demonstrated that a thin delta wing extending along the fuselage could not only provide efficient supersonic flight, but at a suitably high angle of attack (the angle that air flows over the wings), it also could provide enough lift for take-off and landing without resorting to excessively high landing speeds. Her calculations showed that shaping the wing to allow the

---

1 Pevsner, Nikolaus, *The Englishness of English Art*, Architectural Press, London (1956)

formation of vortices *at* the leading edge rather than above or below it would provide enough lift. This was the breakthrough that not only enabled the building of Concorde, but it also gave the aircraft its elegant profile. Küchemann and Weber's work was game-changing in the rarefied world of supersonic design.

The ogival or Gothic arch shape of Concorde's wing was a clever solution to a considerable problem. The supersonic aircraft had to have a wide speed range, from taking off at Heathrow, flying at Mach 2 across the Atlantic, then joining a subsonic queue of aircraft circling to land at Kennedy. The wing was a double-delta shape with conical camber, quite curved. It was set about a third of the way up the fuselage, with fairings on the underside leading up to and over the wheel wells.

The slender delta wing could provide lift at quite extreme angles of attack, whereas lesser wings would stall. It did this by generating conical vortexes over the upper surface at high levels of attack. The airspeed in a vortex is high and so the pressure is low. This provides vortex lift. The Concorde passenger felt these vortices in a form of gentle bouncing at a frequency of around two a second. The wing was the cleverest part of Concorde's design, and took up over 5,000 hours of wind-tunnel testing

Johanna Weber never married, and sometimes longed to return to Germany to care for her mother and sister. But she was charmed by her new home: 'English people typically are such a friendly lot to foreigners, certainly the women,' she recalled. 'I was an odd one, but they wanted all to be kind to me.'[2]

If the aerodynamics of the new supersonic airliner were now understood, there still remained the matter of affording it, and so STAC and the British government started to look for partners to build the aircraft. The French were also working on the idea of supersonic transports and their government asked Sud Aviation to go to Bristol

---

2 Obituary, *The Independent* (Sunday 30 November 2014)

to discuss the idea of a partnership. When the French design was unrolled on the table, the Bristol engineers were surprised to see that the French had come up with a remarkably similar aircraft design to theirs, using Küchemann and Weber's delta wing, and they estimated a similar cost.

Kenneth Owen, in his *Concorde: Story of a Supersonic Pioneer*, explains how the original STAC secret report, which was headed 'For UK Eyes Only', had covertly been passed to the French. Sud Aviation made minor changes to the paper and presented it as its own work.[3] This could go some way to explain why a) the designs were so similar and b) how both teams got the cost so astoundingly wrong. This all goes to show the extraordinary ways in which British politicians and decision makers fiddled with the work of scientists.

The original estimate was for £70 million, but once again a supersonic project met unexpected problems and overruns, with the final amount rising to over £2 billion.[4] That's over *28* times more expensive than estimated, a cost overrun that would make even a London builder blush.

This meant that only 20 airframes were built, and only 14 were operated by Air France and British Airways, the only airlines to buy Concorde. As a result, ticket prices between Europe and the US were around 30 times higher than a budget ticket: a round-trip ticket price from London to New York would cost in today's money £10,000, or $13,000. So the only people able to fly as a passenger in Concorde were the rich and famous.

3 Owen, Kenneth, *Concorde: Story of a Supersonic Pioneer*, Science Museum, London (2001)

4 'The programme's cost, through March 1976, was put at between 1.5 and 2.1 billion in 1976 pounds sterling, or between 3.6 and 5.1 billion in 1977 U.S. dollars...' *New Design Concepts for High-Speed Air Transport*, edited by H. Sobieczky (1997)

Much has been written about Concorde, but maybe not so much about the supersonic airliner's jet engines. They enabled Concorde to fly so fast it made other airliners look slow, as Concorde pilot John Hutchinson observed:

> The only thing that tells you that you're moving is that occasionally when you're flying over the subsonic aeroplanes you can see all these 747s 20,000ft below you almost appearing to go backwards, I mean you are going 800 miles an hour or thereabouts faster than they are. The aeroplane was an absolute delight to fly, it handled beautifully. And remember we are talking about an aeroplane that was being designed in the late 1950s – mid 1960s. I think it's absolutely amazing and here we are, now in the 21st century, and it remains unique.

Supersonic engines for civil airliners were a new territory. How could a jet engine possibly swallow air travelling at over Mach 1?

As is so often the case, it was science-fiction writers who proposed, leaving the engineers to dispose and try to build the damn things. Arthur C. Clarke pointed out that it was Cyrano de Bergerac who had first suggested rockets to go to the moon. He also invented the ramjet engine to get them into space. De Bergerac was the infamous large-nosed dueller, gambler, novelist, playwright and drunkard who was immortalised in the play *Cyrano de Bergerac*, in which the nasally challenged Cyrano agrees to woo the gorgeous Roxane on behalf of his handsome but dim friend.

In his novels *L'Autre Monde: ou les États et Empires de la Lune* (*Comical History of the States and Empires of the Moon*, published after his early and violent death in 1655) and *Les États et Empires du Soleil* (*The States and Empires of the Sun*, 1662) de Bergerac describes a ramjet engine thus:

> I foresaw very well, that the vacuity that would happen in the icosahedron, by reason of the sunbeams, united by the concave

glasses, would, to fill up the space, attract a great abundance of air, whereby my box would be carried up; and that proportionable as I mounted, the rushing wind that should force it through the hole, could not rise to the roof, but that furiously penetrating the machine, it must needs force it upon high.[5]

Ram jets are even simpler than Whittle's turbojet with its one moving part: they have *no* moving parts. Even less is even more. Essentially, they resemble a flying stovepipe in that they consist of a metal tube with a cone-shaped 'centre body' suspended in the mouth, and the addition of some form of fuel. The only problem is that a ramjet prefers its incoming air to be travelling at over Mach 1, so a rocket-assisted launch is necessary.

Approaching the mouth of the tube, the incoming air is compressed by the constricting cone (thereby 'furiously penetrating the machine'), and the air decelerates to subsonic velocity due to shockwaves forming in front of the cone. Then fuel is added to heat the compressed air, just as it is in a turbojet. The expanding gas then exits the ramjet nozzle at high speed. The faster the flying stovepipe goes, the more power is produced, rather like a turbocharged piston engine, so the ramjet tends to run away and accelerates to higher and higher flying speeds. This can endanger the attached airframe unless the fuel supply is controlled.

It is unclear whether he got the idea from his countryman Cyrano de Bergerac, but it was the French inventor René Lorin who properly invented the ramjet as we know it, and as a result was granted a patent in 1913. Then a Hungarian inventor, Albert Fonó, came up with a ramjet fitted into an artillery shell. This would be fired from a conventional gun, then at supersonic speed the ramjet engine would

---

5 From Clarke, Arthur, *Greetings, Carbon-Based Bipeds*, St. Martin's Press, New York (2000)

cut in, propelling the shell as far as 40km. This would vastly increase the range.

It was realised that these new engines could also run at subsonic speeds, given some fettling. As usual, it was the Soviets who got there first, attaching two ramjets to an old Polikarpov I-15 biplane, which briefly became the first ramjet-powered aircraft.[6]

The Germans were next, with Hellmuth Walter building a ramjet, which he tested by bolting it to a lorry, surely another first (and probably last) in world transport. With petrol becoming scarce in wartime Germany, Walter had to use chunks of compressed coaldust in his engine, which was not a success. He might as well have used dried camel dung[7] from Berlin Zoo, as it has an energy density of 15.5MJ/kg – about the same as lignite coal. Both are considerably more energy dense than the best modern lithium batteries at 9.0MJ/kg, which explains the deplorable range of electric cars and aircraft. I take my hat off to the technical virtuosity of the Germans in the 1930s. This was surely due to their technical education system.

Much work was done to perfect the shape of the ramjet centre body, which ended up looking rather like two slender cones joined at their widest point. Moving this centre body forwards and backwards at different airspeeds enabled a plane shockwave to develop in the optimum position: just within the throat of the inlet.

The Americans built a series of ramjet-powered anti-aircraft missiles in the 1950s and '60s. A Talos surface-to-air missile mounted on USS *Long Beach* shot down a Vietnamese MiG fighter in 1968.

---

6 But it was not the last Russian jet biplane. The extraordinary-looking Soviet PZL M-15 jet biplane was used for crop dusting and spraying. This is claimed to be the world's slowest jet at 120mph (200m/h, 110 knots)

7 German tank drivers drove over fresh camel dung in the Western desert, reasoning that it indicated an absence of land mines. The British then planted mines that looked like camel dung. When the Germans realised this, they began to avoid fresh piles of camel manure. The British then made mines that looked like camel dung that had *already* been run over by a tank. Camel dung mixed with Worcestershire sauce provided camouflage for the British tanks, and a tasty snack.

This became the first aircraft to be shot down by a missile fired from a ship. A following MiG was downed after flying through the debris.

Then the British built the Bristol Bloodhound anti-aircraft ramjet, the last line of defence for the V-Force airfields. I remember seeing fields of these olive-green-painted surface-to-air missiles (SAMs) in the 1960s, all pointing towards the east. They were built to protect British V-Force bombers from any Soviet bomber that might sneak past the English Electric Lightning fighters. Bristol acquired the ramjet technology from the Americans but spent a great deal of time perfecting its Thor engine. As a result, the Bloodhound was a great performer, achieving a top speed of Mach 2.7 and a range of 120 miles (190km).

The acceleration of the Bristol Bloodhound made even the Lightning look like a snail: by the time the missile had cleared the launcher, it was already doing 400mph (650km/h). By the time it was 25ft (8m) from the launcher, it had reached the speed of sound (around 720mph). Three seconds after launch, as the four boost rockets fell away, it had reached Mach 2.5 (around 1,800mph). Even so, the wings were made of wood.

The Bloodhound's performance and radar guidance made medium-to-high-altitude flight suicidal for any Soviet intruders, forcing them, like the V-Force, to adopt 'nap-of-the-earth' or hedge-hopping tactics.

Thankfully for everyone, the Soviets never attempted a nuclear attack on the UK, and the Bristol Bloodhound was never used in anger. A planned tactical nuclear Mark III Bloodhound missile was cancelled. This would carry a 6-kiloton nuclear warhead designed to frazzle the electronics of any incoming nuclear missiles with its neutron flux.

Whereas a ramjet decelerates its incoming air to subsonic velocities before combustion by the clever use of shockwaves, a *scramjet* (supersonic combustion **ramjet**) burns its fuel in a supersonic airflow. The air passing through a scramjet remains supersonic all the way through. That allows the scramjet to fly at extremely high speeds, and now we're talking Mach 5. At that speed, a scramjet-powered airliner would be only 90 minutes away from any point on Earth. However, the problems around combustion in the few milliseconds available are immense. Inlet temperatures would be in the order of 1000°C, presenting mind-boggling problems with metallurgy. Scramjets have only recently become capable of flight.

Although it could work well at the speeds Concorde could reach, the ramjet wasn't suitable, as another power plant would be needed to get the aircraft up to the speed at which the ram effect began. Instead, the Concorde's designers used the engines developed for another cancelled British supersonic aircraft, the BAC TSR-2.

The TSR-2 (**T**actical **S**trike and **R**econnaissance Mach **2**) was the planned replacement for the Canberra medium bomber which, in turn, was the jet-engined replacement for the Merlin-engined Mosquito. It was designed for low altitudes and high speeds. Another tragic story of excessive requirements, bickering manufacturers, rising costs, devious politicians and inter-service squabbles consigned the design to the dustbin, but not until hundreds of millions of pounds had been wasted.

The contract had been awarded in 1960 in a shotgun marriage of three companies: the Canberra manufacturer English Electric; Vickers-Armstrongs, the builder of the Valiant V bomber; and the jet engine maker Bristol. The shotgun was held by the British government, which wanted rationalisation of the industry, and the consortium was named the British Aircraft Corporation, or BAC.

Problems started with the three former rivals squaring up to each other. Vickers-Armstrongs was named lead contractor, even though the Valiant had been a failure. English Electric, on the other hand, had built the successful Canberra and Lightning fighter. Meanwhile,

Lord Mountbatten of the Royal Navy constantly interfered, pushing the Navy's subsonic Blackburn Buccaneer.

Bristol too had problems with its engine for the TSR-2, as the Olympus 320-22R jet engines kept blowing up in the test cells. Then a new Labour government was elected, and Harold Wilson and Defence Minister Denis Healey went to the United States, seeking support for a US$8.4 billion loan to Britain from the International Monetary Fund. It is likely that some sort of deal was put together involving the cancellation of the TSR-2 and purchase of the F-111 in return for support over the loan. On the face of it, Denis Healey had good enough reasons: 'The trouble with the TSR-2 was that it tried to combine the most advanced state of every art in every field. The aircraft firms and the RAF were trying to get the Government on the hook and understated the cost. But TSR-2 cost far more than even their private estimates, and so I have no doubt about the decision to cancel.'

The only test pilot to fly the aircraft, Roland Beamont, said, 'The TSR-2 was simply too much for our industry to cope with.'

The F-111's costs escalated even higher than those of the TSR-2 and Britain's purchase of it was cancelled in turn. Eventually, the shorter-range Panavia Tornado, jointly developed by Britain, Germany and Italy, was bought.[8]

The Rolls-Royce RB199 jet engines for this swing-wing aircraft were a joint venture between Rolls-Royce, MTU and FiatAvio, and at first, they suffered from a curious fault. At high altitude and relatively low turbine speeds, the compressor section didn't develop enough pressure to hold back the combustion pressure, and so they tended to backfire (or forwardfire) into the intake, resulting in severe jolting

---

8 A de Havilland BE2 cost £2,068 in 1914, about £240,000 today. A Hawker Hurricane cost £4,000 in 1939, about £265,000 today. One of the last de Havilland Mosquitos in 1951 cost £9,100, now £288,000. It's successor, the TSR-2, in 1965 would have cost £2.6 million each, about *£51 billion* today. And the new F-35 is £70 million, in a programme that will cost $1.5 trillion. Jets aren't cheap.

and vibration. To counter this, the automatic engine controls would increase the minimum idle setting until it was nearly at maximum dry (without reheat) thrust. This is rather like a car engine idling at 6,000rpm and maximum power. As a result, one of the test Tornados got stuck at high altitude at Mach 1.2, with the pilot unable to slow the aircraft down. In the end, he had to resort to a long, hard turn to get the aircraft back under control.[9]

The impact of the TSR.2 cancellation on the UK aerospace industry was immense. Of the 60 firms involved in the project, 57 no longer exist.

The only items rescued from the political wreckage of the TSR-2 were the Olympus 22R engines. These had been designed for sustained flight at over Mach 2.2 for 45 minutes. To provide maximum power over the entire speed range, the inlets for these engines had moveable half-cone shock diffusers with a semi-circular shape.

The first Olympus 22R that was ready for flight by February 1962 was mounted inside Vulcan XA894's bomb bay. It was so powerful it could keep the Vulcan flying without using its own four engines. It was almost too powerful; the Bristol engineers were pushing the limits due to the high performance demanded by the Air Ministry specification. As a result, development had been fraught with problems: two prototype Olympus 22Rs for the TSR-2 exploded during testing, one destroying the Vulcan testbed airframe it was mounted in.

Fortunately, this happened on the ground at Filton: during a full-power run on 3 December 1962, a low-pressure shaft failed. Alan Baxter was there:

---

9 Quinn, Jim, *The White Tornado*, Xlibris, USA (2018)

At about three o'clock power was increased to maximum reheat, the exhaust flame very evident in the grey winter light with evenly spaced shock diamonds clearly visible along its length. Seconds later there was a burst of orange flame above and below the aircraft, which visibly lurched forwards. Inside the cabin the four occupants clearly felt the shock as the engine failed... The LP turbine disc had been ejected from the engine. The disc had apparently travelled around the strengthened bomb bay structure, carving open two 150-gallon stainless steel fuel tanks as it did so. It then hit the ground, and presumably at that point shed what remained of the turbine blades, which must have sprayed like shrapnel into the starboard wing, rupturing every fuel tank. Having bounced off the ground, it then set off in an easterly direction, cutting a ten-foot rip in the port wing leading edge as it went. The marks of its passage were clearly to be seen, savage slashes in the ground every 150 feet or so. It finally stopped not too far from the Bristol T188 (supersonic research aircraft) which was being prepared for a test flight...[10]

The subsequent fire destroyed the Vulcan and a fire engine, although the crews escaped. The cause turned out to be our old enemy: harmonic vibration. This time it was a resonant ringing of the tubular low-pressure shaft, which was being excited to resonance by an air pressure oscillation inside the shaft itself. In the end, though, the Olympus 22R became a powerful and reliable jet engine.

Concorde was built by a consortium of Sud Aviation (later Aérospatiale) and the BAC under international treaty. It had been realised that no one country could afford to build such an advanced aircraft.

---

10 Baxter, Alan, *Olympus: The Inside Story*, Rolls-Royce Heritage Trust, Derby (2007)

A civil version of the Olympus 22R was developed by Bristol for use in Concorde and named the Olympus 593. Bristol Siddeley of the UK and Snecma Moteurs of France were to share the project. At first the design team had considered the use of turbofan engines, but the lower frontal cross-sectional area of turbojets was crucial to supersonic speeds.

The Concorde's engines were unique among civil airliner engines in being fitted with afterburners. These were needed because during design the airframe grew heavier and heavier until there was a 20 per cent shortfall of power for take-off. Perhaps mindful of the fate of the underpowered Comet, an afterburner was added at the back end of each engine.

At Mach 2 the incoming air was compressed and heated up to 120°C by the sheer energy of forward motion. To cope with the high temperatures, the first low-pressure compressor blades were made out of titanium, while those of the last four high-pressure compressor stages were made of our old friend Nimonic 90. The use of nickel alloy here in the compressor is an indication of the thermal stresses involved. The turbine blades were also of Nimonic, and all the blades were air-cooled.

As we have seen, the TSR-2 had used a half-cone centre body in the inlet of the Olympus to slow down the incoming air to subsonic speeds by the cunning employment of shockwaves. Concorde went one better by the use of a variable-geometry flaps in the inlet, which generated shockwaves in the incoming airstream during the supersonic cruise. At take-off, these were opened fully. During one test flight an Olympus swallowed a Concorde intake ramp at Mach 2, suffered 100 per cent compressor damage and still managed 90 per cent of power.

As a result of all this work, the overall pressure ratio for the Olympus 593 cruising at Mach 2.0 and at 50,000ft was no less than 82:1! Compare that to Whittle's first jet engine's ratio of 4:1. The clever intake contributed a pressure ratio of 7.3:1 and the compressors increased that figure again by 11.3:1. As a result, Concorde's Olympus engines were remarkably efficient, recording an overall thermal efficiency in supersonic cruising flight of around 43 per cent – the highest figure recorded for any thermodynamic machine.

Concorde went on to give the fleeing Jet Set more destinations where they could evade the hoi polloi, such as Rio and Bahrain. From rock stars to royalty, Concorde became the only way to travel. But the sonic boom and a disastrous crash in Paris ended the dream. Was the 'World's Most Beautiful Jet' an evolutionary dead-end, or a plane too soon?

It should be pointed out that once again the Soviets were first into the air with a supersonic airliner, the Tupolev Tu-144, or Concordski.

Aleksei Poukhov, one of the Tupolev's designers, said:

> For the Soviet Union to allow the West to get ahead and leave it behind at that time was quite unthinkable. We not only had to prevent the West from getting ahead, but had to compete and leapfrog them, if necessary. This was the task Khrushchev set us... We knew that when Concorde's maiden flight had been set for February or March 1969, we would have to get our aircraft up and flying by the end of 1968.

And they did, just – on 31 December 1968.

The problem was that the Tu-144's development was rushed, and a number of failings emerged. The wing shape wasn't as sophisticated as Concorde's, resulting in terrifyingly high landing speeds of over 200mph 9320km/h). Concordski was unique among commercial airliners in having a braking parachute and ejector seats, but only for the crew (Ryanair might be interested in this concept).

The Tu-144 wing was an ogival delta without conical camber, a cranked delta, set lower on the fuselage than Concorde's wing. It was a simple design and a poor performer at low speed. To give it enough lift, the Tu-144 needed a pair of moustache canards added on – extra retractable wings mounted behind the cockpit. These exacted a weight penalty.

One popular conspiracy theory is that the Concorde design team knew very well that the Soviets intended to steal their blueprints and so they fed them with a convincing but deliberately flawed design. The imprisonment of MI6 courier Greville Wynne at around this time has been produced as evidence of this.

The first engines fitted were Kuznetsov NK-144 turbofans with afterburners, which were needed not only to help it take off but also to maintain it in supersonic cruise, whereas Concorde's afterburners were only needed for take-off and to get through the high-drag transonic barrier. As a result, the Tupolev's range was only 2,500km compared to Concorde's 6,470km.

There were only 55 passenger flights before the Tu-144 was permanently grounded. Interior noise at 90 to 95dB was so high that passengers couldn't hear each other screaming. And this is what they were likely to being doing during one of the 226 in-flight faults clocked up by the Tu-144. Some of these were serious, such as when one of the engines blew up in flight in 1980, with the escaping compressor disc sawing its way through the airframe.

But the most horrifying crash was at the Paris Air Show of 1973, when a Tu-144 sought to out-do Concorde's flight demonstration. The flight crew were under instructions to outperform Concorde's display at all costs.

'Just wait until you see us fly,' said Mikhail Kozlov, the Soviet pilot, to his audience, 'then you'll see something.'

They certainly did. The aircraft departed from the agreed flight profile by making a landing approach with its landing gear and canards extended, then performed a full-power steep climb, whereupon it promptly stalled at 2,000ft and pitched over.

Entering a violent dive with the engines still at full power, the pilot appeared to be attempting to recover control. But the port wing broke up, followed by the rest of the airframe and the Tu-144 hit the ground, flattening 15 houses and killing all six people on board the Tu-144. Eight more died on the ground, including three children. Sixty onlookers were seriously injured.

The Russian theory for the crash was that a French Mirage jet was buzzing the Tu-144, trying to photograph the secret canards, and that Kozlov was surprised into a disastrous evasive manoeuvre. Bob Hoover, an American supersonic pilot disagreed: 'That day, the Concorde went first, and after the pilot performed a high-speed flyby, he pulled up steeply and climbed to approximately 10,000 feet before levelling off. When the Tu-144 pilot performed the same manoeuvre, he pulled the nose up so steeply I didn't believe he could possibly recover.'[11]

Why did Concorde cost so much? Glorious as it was, in retrospect it now looks like another consequence of the Brabazon committee's curious obsession with providing jet-engined transport for small numbers of wealthy and powerful people. One of the many ironies around the aircraft was the fact that Concorde's vast expense was underwritten by the Labour Minister for Technology, Anthony Wedgwood Benn. He was a determined socialist whose political commitment owed more to the teaching of Jesus than the writing of Marx. Jesus told us, 'It is easier for a camel to pass through the eye of a needle than for a rich man to enter the kingdom of God,' so perhaps Benn felt that the rich needed a bit of help.

More significantly, Prime Minister Harold Wilson's 1963 Labour party speech had warned that if the country was to prosper, a 'new

11 Zarakohvich, Yuri, 'The Concordski TU-144: A blast from the past', *TIME* (2000)

Britain' would need to be forged in the 'white heat of a scientific revolution'. This marked Wilson's entry into a debate about the role of science in public life. In 1959, the novelist C. P. Snow had warned that British social and political elites were dominated by 'natural Luddites', whose ignorance of science and engineering made them unfit to govern a world in which technology was becoming ever more important. Tony Benn agreed and saw Concorde as a cutting-edge new technology that could be forged in that same white heat.

The chequebook was open; the requirements of the two nations escalated the expense and the costs of developing a completely new technology overwhelmed the project. Eventually, the British Conservative government of 1984 grew tired of the unending expense and offered British Airways (BA) the Concorde project outright for £16.5 million. As part of the deal, the government threw in two Concordes for BA for just £1 each.

However, operating and maintaining the aircraft was ruinously expensive, and BA and Air France eventually announced on 10 April 2003 that they would be retiring their BAC/Aérospatiale Concordes by the end of the year. The reasons included the crash of Air France 4590 in July 2000 and the post-9/11 downturn in the travel industry.

Sir Richard Branson saw this as an opportunity to offer supersonic travel on Virgin Atlantic. He offered BA 'the same price that they were given them for – one pound – for each of its Concorde aircraft'.

BA declined his offer and retired the fleet.

Does any of this matter now?

'The quality will remain long after the price is forgotten,' said Henry Royce, and years after the crippling expense has long gone, Concorde is still loved and admired.

Meanwhile, Boeing were about to demonstrate once again that it really understood the way the travel market was going. And once again, it was going to bet the company on it.

Chapter 18

747

'Ladies and gentlemen, this is your captain speaking. We have a small problem. All four engines have stopped. We are doing our damnedest to get them going again. I trust you are not in too much distress.'

The speaker was Captain Eric Moody, pilot of G-BDXH *City of Edinburgh*, a Boeing 747-200 Jumbo Jet that was about to win the record for performing the world's longest glide – in a non-glider.

Passengers first noticed a problem with the Boeing's Rolls-Royce RB211-52 engines at 37,000ft above the Indian Ocean on that night in June 1982. Betty Tootell, one of the passengers on the left side of the plane, had a clear view of the port wing: 'To my surprise, it was covered in a brilliant, shimmering light,' she remembered. 'I carried on reading, but I found that I kept reading the same paragraph over and over. I then noticed that thick smoke was pouring into the cabin through the vents above the windows. I didn't know what was happening.'

Other passengers saw a strange blue light shining forwards out of the jet engine inlets, flashing like a stroboscope. Those towards the rear of the cabin saw 30ft flames shooting from the rear of each engine. Shortly afterwards, engine four flamed out and stopped, followed by engine two. Then engines one and three stopped.

Up in the cockpit, there was consternation. The flight engineer, Barry Townley-Freeman, shouted, 'Engine failure number 4!' Then, 'Engine failure number 2!' quickly followed by, 'Three's gone!' And then, 'They've all gone!' Within 90 seconds, they'd lost all four engines.

Up until that point, Flight 009 had been fairly uneventful. The BA scheduled flight had taken off from Heathrow Airport near London on 24 June 1982. After check-in, the 263 passengers had settled into their seats, ordered drinks from the cabin crew, lit up cigarettes and prepared for the flight that would take them to New Zealand via India, Malaysia and Australia.

The first intimation of disaster was a blue flickering light on the windscreen, similar to St Elmo's Fire. Then, back in the passenger cabin, the chief steward, Graham Skinner, smelled smoke. In 1982, it was still legal to smoke on jets, with up to a third of passengers puffing away happily. Skinner thought that there was perhaps a smouldering cigarette in a bin. But this smoke smelled of sulphur...

Up in the cockpit, Captain Moody had just returned from a lavatory break downstairs. 'The smoke filling the plane smelt like a sulphuric, electrical smell,' he remembered. 'I went on the flight deck expecting to hear that we had some electrical smoke from the aircraft.'

Staring out into the night ahead of them, the crew saw they were flying into a sheet of brilliant white light, and the temperature within the aircraft began to rise. They could not understand what was happening.

Back in the passenger cabin, Skinner was suffering: 'It got really, really hot. You were perspiring, drenched in sweat. The acrid smoke filling the cabin was at the back of your throat, up your nose, in your eyes – your eyes were running.'

The crew were well prepared to cope with disaster that night. Captain Moody's first flight at the age of 16 had been at the controls of a glider, and he was one of the first pilots trained on the Boeing 747. Roger Greaves, the first officer, had been a co-pilot for six years, and the alert Barry Townley-Freeman was a flight engineer. The crew had just taken over the aircraft at Kuala Lumpur in Malaysia, and they were well rested.

Captain Moody now realised he was at the controls of the biggest and heaviest glider in history. He knew that his 747-200 had a glide ratio of approximately 15:1, so, at best, he was able to fly 15km for each 1,000m the Jumbo descended. First Officer Greaves calculated they had less than 25 minutes and 90 miles (145km) to go before the aircraft hit the ground, or the sea. And the flight crew still had no idea what the problem was. Had they run out of fuel? Was there something wrong with the air?

Their first job was to radio Jakarta control: 'Mayday, Mayday. Jakarta control. Speedbird nine. We have lost all four engines. Repeat, all four engines. Now descending through flight level 3-5-0.'

This was an unusual message and Jakarta at first mis-heard it as 'We have lost engine four.'

Now the crew began the standard engine restart drill and carefully turned the Boeing back towards the nearest possible airport, Jakarta. But Greaves worked out they would not be able to make it without at least one working Rolls-Royce RB-211. Then the cabin pressure began to fall, and oxygen masks fell from the passenger cabin ceiling. But First Officer Greaves' mask was defective and useless, as the delivery tube had detached itself from the mask.

Now Captain Moody had no engines, and soon, no first officer. He took a quick decision, which probably saved all their lives. He shoved the Boeing's nose down and went into a steep nosedive, losing 6,000ft in one minute to an altitude at which his passengers and crew could survive. This action lost precious height, and now at 13,500ft, the Jumbo was approaching the altitude at which it would have to turn back over the ocean and attempt a ditching. No one had ever tried this before in a Boeing 747. The crew looked at each other.

Then, as the Jumbo fell out of the sky past 13,000ft, during Townley-Freeman's increasingly frantic attempts to restart the engines, something happened. Rolls-Royce engine number four choked, gulped, spat something out and roared back into life. Moody used that engine's power to reduce the rate of descent. Shortly thereafter, engine three restarted, allowing the plane to climb slowly.

Shortly after that, engines one and two successfully restarted as well. The crew then asked Jakarta control for permission to climb high enough to avoid crashing into Indonesia's mountains.

Their troubles were not yet over. During the approach engine two coughed and stopped again. As the airport came within view, the crew realised they couldn't see anything out of the windscreen, which looked as if it had been sandblasted. Captain Moody decided to do the final approach on the instrument landing system, but now they found out that Jakarta Airport's vertical guidance system was broken. The landing lights on the Jumbo didn't seem to be working either. The crew only had the airport's lateral guidance as a guide. First Officer Greaves had to monitor the airport's distance measuring equipment (DME), then called out how high they should be at each DME step along the final approach to the runway, creating a virtual glide slope for them to follow.

Captain Moody executed a perfect landing. But then he found it impossible to taxi because the glare from the apron floodlights made the already sandblasted windscreen completely opaque.

What had happened to make all four engines shut down? The flight crew initially blamed themselves for missing something, but after Rolls-Royce engineers stripped down the RB211s, they found the cause: they were clogged with ash. The 30ft-long flames that passengers had reported were backfires caused by too high a fuel-to-air ratio due to the strangling of the airflow.

It turned out that there had been an eruption of the Mount Galunggung volcano southeast of Jakarta that day. The Boeing 747 had flown into a cloud of volcanic ash, and this had not only sandblasted the aircraft but also choked its engines. The cloud didn't show up on radar screens because volcanic ash is a dry material, and rain clouds are detected by their water particles.

In the end, the plane had been saved by exceptional flying skills from the crew and the sheer luck of a broken oxygen mask. The sharp descent into clean air seemed to have shaken the volcanic slag out of the engines and allowed them to restart. A long, slow, gentle glide to the sea may not have provided the requisite conditions.

The Rolls-Royce RB-211 might have been able to swallow volcanic ash but it couldn't swallow a chicken. This inability ultimately brought Britain's finest company to its knees and Rolls-Royce was nationalised as a result.

The RB-211 engine was a high-bypass turbofan with no less than three sets of turbines spinning three concentric shafts within one another. These drove three sets of compressor blades all running at their optimum speed. This extra complexity gave greater efficiency leading to lower fuel consumption and reduced noise.[1] Whittle's 'one moving part' seemed a long time ago now. To save weight and expense Hyfil carbon-fibre compressor blades were used instead of titanium.

After passing all the other tests, the RB211 faced the bird-ingestion test. This airworthiness requirement was for the engine to ingest eight 2.5lb (1.1kg) birds at the forward speed of the aircraft at rotation. After bird ingestion the engine had to keep delivering 75 per cent of take-off power for 30 minutes to allow the aircraft to land.

A bird gun, a large-diameter compressed-air cannon, was loaded with a chicken, and it fired the dead bird into the RB-211 test engine. The carbon-fibre blades shattered at the impact.

Rolls-Royce had understood the risks of Hyfil and had developed a titanium blade in parallel as insurance, but this meant extra cost and more weight. Then it discovered that only one side of the titanium billet was of the right metallurgical quality for blade fabrication.

---

1 It is beyond the scope of this book to detail the immense complexity of the RB-211. For more, see Ruffles, Philip, *The History of the Rolls-Royce RB211 Turbofan Engine*, Rolls-Royce Heritage Trust, Derby (2014)

The resulting overruns and extra expense were enough to double the original development budget to £170.3 million, and the production cost of each engine rose to £230,375, more than the selling price. Rolls-Royce declared insolvency in January 1971 and the company was nationalised by the government, a humiliating end for a proud company. Frank Whittle bitterly criticised the nationalisation of jet propulsion development, saying it was responsible for Britain losing its world lead in jet engine technology.

Then things improved for the RB-211. The brilliant Stanley Hooker was persuaded to return from retirement, and the old gang – Hooker, Cyril Lovesey and Arthur Rubbra – fixed the problems on the engine, which was finally certified on 14 April 1972. Hooker was knighted for his role in 1974.

The engine turned out to fulfil its promise, as it became reliable, powerful and efficient. It was found that BA's Boeing 747s fitted with RB.211s burned roughly 7 per cent less fuel than its Pratt & Whitney JT9D-equipped fleet, a saving of about $3.155 million a year per aircraft at today's prices.

The Rolls-Royce RB-211 was fitted on many aircraft, even the Russian Tupolev Tu-204-120 airliner, the first fitment of Western jet engines to a Russian airliner.

Perhaps they remembered the Rolls-Royce Nene.

Once again it was Pan Am's president, Juan Trippe, who influenced the future of long-distance jet travel. In the early 1960s, he asked Bill Allen of Boeing to build an even bigger airliner than the 707, in fact he wanted one twice the size.

The success of the jet travel market had led to increasing congestion at airports. Too many passengers were being carried by aircraft that were too small. Trippe realised the solution was, once again, to pile them high and sell 'em cheap. But to do that he needed a *really* big jet airliner.

'If you build it, I'll buy it,' said Trippe during their Alaskan fishing trip, to which Allen replied, 'If you buy it, I'll build it.'

And so a Jumbo was born.

The suggestion came at a good time. Boeing had just completed a study for the CX-HLS, a new giant cargo plane for the USAF. The CX-HLS had to be loaded from the front, with a huge front door where a cockpit usually was. This necessitated a cockpit mounted on top of the fuselage, blending into a top cabin that ended by the wing. This giant new plane would demand engines that were more powerful and more economical than anything that went before, and General Electric, Curtiss-Wright, and Pratt & Whitney all worked on the idea and submitted engine proposals.

In 1965, the CX-HLS contract was won by Lockheed's design for the new C-5 Galaxy transport, using General Electric engines. But Boeing's work did not go to waste. It used its distinctive upper-deck design for the new 747. And so, on 9 February 1969, model No. 1 of the 747 took off from Everett's Paine Field. It was the largest commercial airplane in the world, capable of carrying more than 400 passengers. There were minor teething troubles, such as the vibrations during the take-off roll that caused over-stuffed luggage bins to burst open, but on the whole the airframe behaved well, considering its size.

Pan Am, of course, were the first airline to fly the 747. 'The first one has our name on it... where the big thing is comfort,' said the ads. 'With two aisles throughout. A double-decker section up front, complete with upstairs lounge... And three living room-size Economy sections... And seats almost as big as First Class... you want to fly the plane that's a ship, the ship that's a plane.'

For Americans, the Boeing 747 had a particular significance.

'The late 1960s were a time of real turmoil,' said Boeing historian Mike Lombardi. 'We had the Vietnam War, we had the race problem, people in the streets rioting. Then two things happened: the 747 and the Apollo program, going to the moon. Both showed that despite all the differences and turmoil, we could come together and do something great. Boeing had built the world's biggest airplane. It's still part of the DNA of the company.[2]

The giant aircraft demanded giant engines. The Pratt & Whitney JT9D engine was a high-bypass-ratio jet engine, and once again development was fraught with problems. When taking off with the full 43,500lbf (193kN), the turbine blades rubbed against the sides of the engine and self-destructed. But back on the ground everything seemed to have the correct clearances. It was eventually realised that the whole engine was turning oval under the stress of full-power take-offs, and the engine housings had to be strengthened as a result. These engines would eventually develop 50,000lbf (220kN), and the General Electric GEnx-2B67 jet engines fitted in the last 747s would develop no less than 66,500lbf (296kN).

If they weren't stopped by a cloud of volcanic ash.

What else can stop your engines? Well, running out of fuel will do it. In the infamous 'Gimli glider' event in 1983, a nearly new Air Canada Boeing 727 suffered a double engine stoppage at 41,000ft over Ontario. Most of the instruments went blank, too. Hastily looking through the emergency handbook for a section entitled 'What to do when both engines fail', the pilots, Captain Pearson and First Officer Maurice Quintal, found there was no such section.

The Boeing 727 is an automated aircraft with every instrument and function controlled by computer. Computers, of course, rely on electricity to work. This is usually provided by the two Pratt & Whitney turbofan engines.[3] Without fuel though, the engines gave nothing: no power and no electricity. They were just expensive

---

2 *Seattle Times* (9 February 2019)

3 With 48,000lb of thrust each, giving a cruising speed of Mach 0.80 or about 480 knots (552 mph), and a 5,000-mile range. When they have fuel.

airbrakes. All attempts to restart failed, and the lack of engine power meant there was no electrical power for the electronic flight instrument system and little hydraulic power for the flying controls. They now had fewer working instruments than a 1930s biplane, and less engine power.

The crew then deployed the ram air turbine (RAT), which lowers an impellor into the airstream. This drives a small backup generator, which provided just enough power to operate the control surfaces of elevator, rudder and ailerons.

After a 45-mile glide they approached their emergency landing strip, an ex-Royal Canadian Air Force base at Gimli, Manitoba. As they approached, they found there was no electrical power for extending flaps, which would have slowed the aircraft down and provided it with more lift, and no power to deploy air brakes or even to lower the landing gear. Again, there was nothing about this in the handbook. Taking a guess, Pearson pulled a gear-extension switch. The heavy main wheel assembly dropped into the down-and-locked position, but the lighter nose gear only partially deployed and didn't lock down. This probably saved the day.

Luckily, Captain Bob Pearson, like Captain Moody, had trained as a glider pilot. Approaching his emergency landing strip too high and too fast, he lost altitude by putting in a rare manoeuvre for a jet passenger airliner: a 'forward slip'. This is performed by applying the rudder in one direction and the ailerons in the opposite direction. This is called 'crossing the controls'. With consummate skill he put the Boeing down on the centreline of the runway.

Unfortunately, on that day the decommissioned airstrip was being used as a racetrack by the Winnipeg Sports Car Club. The silent arrival of a large airliner was an unexpected bonus for the audience. The first thing they heard was the crash of the collapsing nosewheel landing gear, and three boys bicycling along the runway had to pedal for their lives. Art Zuke was one of them: 'I saw this thing flying sort of sideways and cockeyed.' Zuke's observation was spot on – he saw and described Pearson's forward slip. Due to the collapsed nose gear, the aircraft ground to a halt just before piling into the amazed spectators. Some of them thought it was all part of the entertainment and applauded.

Investigation revealed that all three fuel gauges were blank and not working when the aircraft was taken over. The amount of fuel had not been checked with measuring sticks, as regulations stated. Furthermore, there was a miscalculation of pounds to kilograms of fuel by mechanics, fuel servicing personnel and co-pilot Quintal. This led to only to half the fuel needed being pumped aboard. All in all, there was a concatenation of faults, assumptions, miscommunication and general sloppiness. The final accident report read:

> There would have been no accident on July 23, 1983… had the fuel processor on aircraft 604 not malfunctioned. The human errors would never have occurred had the fuel processor that operates the fuel gauges been functioning properly. Nor would the shortcomings of Air Canada, Transport Canada, and the fuelling companies have been exposed.

This all seems extraordinary. Would you hire a car with a broken fuel gauge? And if you did, wouldn't you double-check the quantity of fuel you put in it? And, amazingly, this multi-million-dollar aircraft also lacked a reserve fuel tank, a feature found on 1960s Triumph Herald cars and assorted mopeds.[4]

In the end, though, it was Captain Pearson's old-fashioned rudder-and-stick skills that saved the lives of his passengers and

---

4 Until the mid-1960s, VW cars didn't have fuel gauges either, but they did have a fuel reserve tap, which stuck up near the driver's feet. This had to be frantically operated with your toes when the Beetle ran out of fuel in the fast lane of the autobahn – but only if you had remembered to turn it back to 'main tank' the last time you ran out…

the aircraft. He and his co-pilot were awarded with the Diploma for Outstanding Airmanship by the FAI (World Air Sports Federation).

The Boeing 747 gave people like me cheap long-distance travel, something I will always be grateful for. I just wished I could have afforded to travel in the beds that were eventually provided for first-class passengers. But I believe I was a pioneer in this field in a small way.

I'm one of those people who just cannot get comfortable in an airline seat. You'll find me sprawled asleep on the galley floor, balanced across three empty seats or, when the plane is full, just staring morosely at the seat ahead of me.

Flying out to Southeast Asia in 1998, I found myself on one of the night flights sitting bolt upright in cattle-class, as usual, bitterly wishing that I could have afforded a less knee-crushing seat. After hours of agony, I got up and wandered around the back of the Boeing 747. There was little going on. Most of the passengers were sprawled across one another, pretending to sleep. Then I saw it. A ladder, right next to the lavatory door on the starboard side. And a notice, in red: Crew Only.

It was irresistible. I crept up the ladder, and gently raised the trapdoor in the ceiling. As it opened, I had a brief vision of being sucked out into a star-lit night at 36,000ft – but no. As my eyes slowly took in the scene, I became aware of a little bedroom, tucked under the tail plane. There were eight bunks, each with its own night-light and safety belt. And they were all empty...

As I crept back to my seat, my mind was working overtime. I knew that some long-haul flights carried spare crews, but I assumed that they slept in seats (or rather, didn't sleep). This was obviously a high-altitude bedroom. And on the return flight, I thought, I would get myself up there – obviously they didn't use it on this sector. And if they couldn't supply seats big enough to sleep in, why shouldn't I get the chance to relax up there?

After a trip around Sabah, pursuing bird's-nest soup, I found myself back on the same 747 returning to London. We took off, the meal came around and night fell. One by one, my fellow passengers adopted positions reminiscent of the occupants of an electric chair. One by one, their lights went out. And then I crept away.

It went like clockwork. A feint towards the starboard lavatory. A quick shuffle up the ladder, lift hatch, check for inhabitants – and in.

I selected a comfy looking bunk at floor level, crawled in and lay down. Oh, bliss! Oh, what luxury! I clipped my safety belt on and wondered why on Earth couldn't we all travel like this? If, instead of a row of seats, we could have a stack of box-bunks like long pigeon-holes, what comfort we could travel in! Safe in our long aluminium box, we could withstand accidents better, sleep better, travel more.

I began to notice a strange thing. At this end of the 747, the fuselage was wagging slowly and rhythmically, rather like the tail of a sleepy fish. And, peering through a crack in the bulkhead I could see all the way along the roof space above the passengers' ceiling. There were lights in there, and the sides were heavily insulated. I fell asleep.

I awoke with a start. Something was going on. I peered over and saw the hatch being lifted. A head appeared. I turned to the wall and froze. Bloody idiot! Why did I assume no one would come up!

The pilot came up the ladder, and for a long time I felt his eyes boring into the back of my neck. I tried to look like an air hostess, something I don't have much of a talent for. Then I felt him look away, and after an interminable time I heard him get into his bunk. Then, one by one, three other crew members came up and, one by one, I could feel them examining me closely.

It was awful. I was in a cold sweat of fear. I was going to be a named hijacker. I was going to be dumped at an Iraqi airport and handed over to the counter-terrorism squad. I was rigid with horror.

One by one, my bedfellows rolled over and pulled up their blankets. And, one by one, their little night-lights went out. Eventually, silence fell; as well as it can fall behind the tail pipes of four giant Pratt & Whitney JT9D-7 jet engines. But were the crew asleep?

I had to get out. I must have been lying there for an hour wondering what to do. I couldn't stay there, as they would drag me out when they got up to take over their shift.

I had to do something. Very, very carefully I undid my seat belt. And very, very carefully I picked up my shoes and very, very carefully I crawled across the floor. As I was halfway across, I had a powerful vision of a torch being snapped on and a cold Biggles-like voice from behind the snout of a pistol: 'Freeze!'

But it didn't happen. In one swift ape-like movement I whipped open the trapdoor and shot down the ladder. I then disappeared into the lavatory and collapsed on to the seat sweating profusely. After a while, I crept back to my seat and settled back into the anonymity of economy. My seat felt almost comfortable. But I didn't sleep.

Years later, I stayed at a Swedish hotel that has been made from an old 747-200. The Jumbo Stay hostel at Stockholm airport gave me a better night's sleep than the flying bedroom.

Covid-19 killed the 747, with BA retiring its 31-strong fleet of 747s in July 2020, blaming the pandemic and a consequent downturn in travel. And Qantas CEO Alan Joyce succinctly explained why the Jumbo had been so important:

It's hard to overstate the impact that the 747 had on aviation and a country as far away as Australia. It replaced the 707, which was a huge leap forward in itself but didn't have the sheer size and scale to lower airfares the way the 747 did. That put international travel within reach of the average Australian and people jumped at the opportunity.

The 6,000-mile range of the 747 revolutionised travel. 'The range of the plane allowed it to go anywhere in the world,' said Michael Lombardi, Boeing's corporate historian. 'It was at that point in history where all of humanity had the ability to get on a flight.'[5]

In the end, it was high-bypass-ratio turbofan jet engines that made the Boeing 747 so successful. As we have seen in this book, the turbofan avoided the compressibility problems of the old propeller-driven aircraft, turbofan engines were much lighter, per unit of power, than piston engines; they developed huge amounts of power in a single unit with mechanical simplicity, and by the 1970s, the high-bypass designs delivered the same overall efficiency as that of the best reciprocating engines. Finally, the turbofan was more reliable than the old piston engines, and that increased the safety of modern airliners.

The year of peak Jumbo was 2002, when 747s of 50 airlines flew 10.5 million passengers on 33,000 flights.

'It was the plane that shrank the world,' said Lombardi. 'That is the legacy of the 747.'

---

5 In fact, only 20 per cent of the world's population has ever set foot in an airliner.

Chapter 19

Brace, Brace, Brace

Have you ever flown across the Atlantic, idly looked out of the window and realised the jet airliner you're sitting in only has two engines?

Have you then wondered what would happen if one of the engines suffered a total failure? Or if *both* failed? Jet engines are far more reliable than piston engines, but, occasionally, they do break down.

The consequences of an airliner ditching in the mid-North Atlantic in winter do not bear thinking about.

But it happens.

In May 1970, a twin-engine Douglas DC-9 ran out of fuel and ditched in the Caribbean Sea 30 miles (48km) off St Croix. The public address system was out of action, so the pilot could not warn of ditching, and, therefore, many of the passengers were standing or had their seat belts unfastened. The sea was rough, but the aircraft remained in one piece. The crash landing ('ditching' is a polite term) resulted in 23 deaths, and injuries to all except three of the 40 survivors. The plane sank soon afterwards. The survivors managed to stay afloat in their life jackets for up to 2½ hours until they were rescued.

Bear in mind this was early summer in the Caribbean, the sea was relatively warm and helicopter rescue came within 1½ hours. None of these conditions would be obtained in the mid-Atlantic, so it can be expected that few would survive such an event.

Can you do anything to help your chances of surviving a ditching? Landing on rivers helps. The famous case of the 'Miracle on the Hudson' wasn't really a miracle. Chesley Sullenberger's twin-engined

Airbus A320 hit a flock of Canada geese on 15 January 2009, while taking off from New York's LaGuardia Airport. Both engines lost power. Unable to reach any airport for an emergency landing, Sullenberger ditched in the Hudson River off midtown Manhattan.

The air stewards shouted 'Brace! Brace! Brace! Heads down!' They described the landing on the water as 'one impact, no bounce, then a gradual deceleration'. Sullenberger had landed near some boats, and all 155 people on board were rescued by the crews on these, with few serious injuries. Due to the pilot's skill and the fortunate watery landing strip, this was probably the most successful ditching in aviation history.

Does bracing before impact work? Most certainly. In January 1989, a twin-engined Boeing 737-400 crashed into a motorway embankment near Kegworth, Leicestershire, in England while trying to make an emergency landing.

A fan blade had broken in the left-hand engine, a CFM International CFM56,[1] filling the aircraft's cabin with smoke. The pilots thought that this indicated a fault with the other engine because earlier versions of the 737 ventilated the cabin from the right-hand engine. So they shut down the one remaining engine and selected full thrust with the damaged engine. This added fuel to the fire, both engines went out and the 737 crashed.

Of the 126 people aboard, 47 died and 74 sustained serious injuries.

A careful study was made, and it was found that those who adopted the correct brace position survived best. Essentially this consists of placing your head on, or as close as possible to, the surface it is most likely to strike; either the seat in front of you or the bulkhead.

---

1 Revised fan blades on the low-pressure compressor stage of the uprated CFM International CFM56 engine vibrated badly when operated at high power above 10,000ft. As it was an upgrade to an existing engine, in-flight testing was not mandatory, and surprisingly the engine had only been tested in the laboratory. Vibration created metal fatigue in these fan blades and on G-OBME this caused one of them to break off, destroying the rest of the engine.

Bending at the waist means that you won't slide under the seat belt ('submarining'), or violently jack-knife. Your body is going to flail like a marionette in a bomb blast, so planting both feet on the floor with a bag between them avoids one leg breaking the other.

There are a couple of internet myths around the brace position. One is that it is only for preserving a corpse's dental integrity for identification after a crash; another is that the position is designed to increase the chance of death (by breaking the neck) to reduce medical insurance pay-outs. Both of these are nonsense, and they are untrue.

We have seen the backwards-facing seats of the proposed Avro Atlantic. In the event of a crash, these seats are safer, but airlines would be unlikely to welcome the idea. 'During an impact, the passenger's centre of gravity would be higher, and the seat would be taking more of the strain,' explained David Learmount, operations and safety editor at the aviation news website FlightGlobal.com. 'Therefore, the seat itself, the fittings and the floor of the aircraft would need to be strengthened. That would increase the weight of the aircraft, which would increase fuel consumption.'

Learmount added that the same weight problem would apply to three-point seatbelts, as the centre of gravity would move from the waist to the shoulder. 'From a safety point of view, they are attractive ideas, but can you imagine an airline like Ryanair supporting it?'

In the pursuit of lightness and fuel economy, Ryanair has now reduced the size of its in-flight magazines, serves less ice and has asked its cabin crew to watch their figures. The Irish airline has even proposed 'standing-room only' cabins, with CEO Michael O'Leary saying, 'If there ever was a crash on an aircraft, God forbid, a seatbelt won't save you.'

This is not entirely true, as it depends on the kind of crash, and 'vertical seating', as it is dubbed, is unlikely to be approved by safety regulators.

Learmount also pointed out that rear-facing seats were not popular with passengers.

'People wouldn't want them,' he said. 'British European Airways used to fly Trident jets with both forward – and rear-facing seats – and people would kill for a seat facing the front. On trains it's always the forward-facing seats that are worn out.'[2]

Are twin-engined jet airliners safe?

The old twin-engined piston aircraft were notoriously hard to fly with one failed engine. The extra rudder angle and increased throttle settings threw more load onto the good engine, which often failed in turn.

As we have seen, the civil aviation safety authorities had a standard named ETOPS: Extended-range Twin-engine Operational Performance Standards. Based on the old piston-engine reliability statistics, the FAA had stipulated that a twin-engine airliner could fly no more than 60 minutes flying time away from the nearest airport equipped to handle an emergency landing.

TWA asked for ETOPS-60 to be relaxed to enable its twin jet-engined Boeing 767s to fly across the Atlantic between Boston and Paris. Millions of dollars were at stake. In 1985, this was granted and dubbed ETOPS-120. This regulation allowed a twin-engine airliner to fly no more than two hours flying time away from the nearest airport equipped to handle an emergency landing. As jet engines became more and more reliable, ETOPS rules were slowly relaxed from 60 minutes to 120 minutes, and then up to 180 minutes.

---

2 *Daily Telegraph* (23 July 2019)

ETOPS-180 was enough for manufacturers to shift their development towards today's cheaper long-range twin-jets.

In 2014, the European Aviation Safety Agency (EASA) approved the new Airbus A350-900 airliner for ETOPS 'beyond 180 minutes' diversion time. This included an option for up to 370 minutes maximum diversion time.

And in December 2019, the EASA gave the Rolls-Royce Trent XWB-97 a maximum diversion duration of 420 minutes for the Trent, including 405 minutes at maximum continuous thrust, plus 15 minutes at hold thrust.

And so the twin-engined jet airliner had, at last, been officially declared safe.

The long career of the four-engined Boeing 747 had encompassed the several decades of improvements to turbofan jet engines. The Rolls-Royce RB-211, Pratt & Whitney JT9D-7 or GE CF6 were fitted to the 747 at first, then the 747-8 had either the Rolls-Royce Trent 1000 or the General Electric GEnx. These new-generation engines had curvy swept composite fan blades that are immediately recognisable by a boarding passenger. But once again, manufacturers had problems with engine blades.

The Rolls-Royce Trent 1000 was one of two engine options for the new Boeing 787 Dreamliner, a wide-bodied jet airliner with an airframe primarily made of composite materials. It only has two engines. The Trent is a three-spool engine with three coaxial shafts running at different speeds and directions of rotation. It has a single annular combustion chamber with 18 high-pressure fuel spraying nozzles. Incoming air arrives first at the huge swept blades of a 9ft 4in (2.85m) fan. This is driven by means of the low-pressure shaft by no less than six axial turbines. The eight intermediate compressors are driven by a single turbine through the intermediate shaft, and the six high-pressure stages are also driven by a single turbine. However, this high-pressure shaft revolves in the opposite direction to the other two. The bypass ratio is 10:1. All this work resulted in 62,264 to 81,028lbf (276.96–360.43kN).

Unfortunately, it was discovered in 2016 that some of the Trents suffered from blade cracking of the intermediate pressure turbine blades. This was due to corrosion related to the high sulphur content in the atmosphere over some Asian cities. The fault grounded up to 44 aircraft and cost Rolls-Royce at least £1.4 billion.

The other engine used in the Dreamliner was General Electric's GEnx, and that had its problems, too. Four 747 freighters suffered from icing in their GEnxs and then, in 2014, a Dreamliner had its first in-flight shutdown when a Japan Airlines (JAL) flight experienced an oil-pressure alert. On 29 January 2016, another JAL Dreamliner had a shutdown after icing up. The General Electric GEnx-1B PIP2 engine suffered damage while flying at an altitude of 20,000ft, and it could not be restarted. The damage was caused by a fan imbalance resulting from fan ice shedding. The plane was able to land safely with its one remaining engine, but the FAA described this as an 'urgent safety issue' and ordered fixes to the engines.

When the limits of engine technology are pushed out so far ahead in an atmosphere of cut-throat competition, things tend to break.

The new twin-engined jet took off from Addis Ababa, the Ethiopian capital, on its way to Nairobi, in Kenya. It was early morning, and all was well. The Boeing climbed steeply.

Then, suddenly, the nose dropped. Mystified, the pilots pulled the nose up, only for the aircraft to take back control, putting its nose down again. The aircraft seemed to have a mind of its own, constantly fighting the pilots when they tried to make it climb. A few seconds later the automatic system forced the nose down again.

They were now getting dangerously close to the ground. An alarm sounded and a metallic voice alarm grated, 'Don't sink! Don't sink! Don't sink!' Captain Yared Getachew felt the control column start to shake in his hands, a signal that his plane was nearing a stall.

He and his first officer, Ahmednur Mohammed Omar, swiftly turned off the electronics and attempted to fly the plane manually. It was too difficult: the speed was now so high that the two men struggled to pull back the control columns. The nose wouldn't come up. Getachew tried switching the electronics on again, the nose came up, then fell again. The aircraft's dive grew steeper. It reached 500mph, then it plunged into the ground.

Five months before, the same kind of accident had happened to another 737 MAX 8 in Indonesia, with the aircraft swooping and falling like a dolphin through the sky before it, too, crashed.

After the investigations were concluded, the finger pointed at a little-discussed piece of flight control software. It was supposed to make the airliner familiar to pilots who were used to the previous models of the Boeing 737. The new 737 MAX had a tendency for the nose to rise too much during take-off, and this software forced the nose back down without informing the pilots. In both crashes, it seemed to have malfunctioned and forced the nose down when the aircraft was trying to climb.

Boeing's consequent grounding of all its 737-Max aircraft has been the biggest crisis in the corporation's century-long history. It will cost in the region of US$20 billion. The only comparable grounding was that of the de Havilland Comet jet airliner.

In total, 346 people lost their lives in the two 737 Max crashes of the Lion Air flight in Indonesia and the Ethiopian Airlines jet in Ethiopia. Most of the media attention has been concentrated on the deficiencies of the flight control software, but further examination reveals the cause of it all: the positioning of its new jet engines.

And why were the engines put where they were? Because of the height of a flight of stairs.

The Boeing 737 first flew in April 1967, back in the Cold War. It had the same fuselage diameter as the 707 and was intended as a commuter jet. In those days before jetways, the 737 had metal stairs that folded down to the ground and passengers climbed up these to board the aircraft. This was before baggage conveyor belts, so the handlers had to heft heavy luggage by hand into the hold from the tarmac. As a result, the 737 had to be close to the ground.

This was acceptable at the time as the aircraft first had Pratt & Whitney JT-8 engines, which had an inner fan diameter of 49.2in. These engines were long and skinny, like cigars, and they fitted beneath the wings. The bottom of the engines was just 17in from the tarmac. Most aircraft now have a clearance of around 29in.

However, as jet engines evolved with more bypass, they became fatter. For the previous 737-300, which had the CFM56-7 engines, Boeing came up with a clever fix: it created a flat bottom on the nacelle (the shroud around the fan), creating what pilots came to call the 'hamster pouch'.

Then, in 2011, Airbus launched the A320neo fitted with more economical CFM LEAP-1A engines and got 667 orders at the Paris Air Show, a record for a commercial aircraft. Worse still, American Airlines ordered 130 of the new Airbus and 130 of the older one. Boeing had to answer back, and fast. But instead of spending money on a new design, it once again modified the old 1960s 737. Industry insiders were surprised.

'Boeing has to sit down and ask itself how long they can keep updating this airplane,' said Douglas Moss, an instructor at the University of Southern California's Viterbi Aviation Safety and Security Program, a former United Airlines captain, an attorney and a former air force test pilot. 'We are getting to the point where legacy features are such a drag on the airplane that we have to go to a clean-sheet airplane.'[3]

But it is much easier and cheaper to build a new derivative of an existing design than attempt a clean-sheet design. Boeing's answer was the 737 MAX, also mounting the fatter CFM LEAP-1b engines. But these engines on the Max 8 have a diameter of 69in, nearly 20in wider than the original. There wouldn't be enough clearance, even with a hamster pouch. How to stop them dragging along the tarmac?

Boeing's solution was to extend the engine-bearing pylons so that the engines were mounted further forward and higher up. The engineers also extended the nose landing gear. Now there was the requisite 17in of ground clearance. But this was a bodge.

This new position for the engines generated extra lift and made the nose of the 737 MAX pitch up in flight. To cope with this handling oddity, Boeing installed flight control software called Manoeuvring Characteristics Augmentation System (MCAS), which instructed the stabilizer on the tail to push down the nose if the angle of attack got too high. The MCAS depends on data from two sensors. But it appears the MCAS relied on a sensor that was wrongly reporting a high angle of attack when the airliner was nowhere near stalling. Despite the pilots desperately trying to fight the MCAS, both 737 MAX airliners crashed.

The September 2020 report by the US House Committee on Transportation and Infrastructure[4] into the Boeing 737 Max found that, in both crashes, a single faulty attack angle sensor triggered the

---

3 *Los Angeles Times* (15 March 2019)

4 Reported by Simon Calder at https://www.independent.co.uk/travel/news-and-advice/boeing-737-max-plane-crash-faa-evidence-b455872.html

software system. MCAS was designed to be pilot proof to the point where the computer could even point the plane into the ground and defeat the strength of the flight crew to overcome it.

The report stated that Boeing knew about this problem, concealing internal test data from 2012 that showed one of the company's test pilots taking more than ten seconds to diagnose and respond to uncommanded MCAS activation in a flight simulator. But Boeing withheld crucial information from the FAA and its airline customers and concealed the existence of MCAS from 737 Max pilots.

The report noted: 'There was tremendous financial pressure on Boeing and the 737 Max programme to compete with Airbus's new A320neo aircraft.' It would cost Boeing around US$1 million per aircraft if pilots required simulator training to convert onto the new 737. And the chief technical pilot boasted of his successful efforts to talk airlines out of the need for simulator training, writing to a Boeing colleague: 'I save this company a sick amount of $$$$.'

Internal messages between other employees revealed their own doubts: 'This airplane is designed by clowns who in turn are supervised by monkeys.' They referred to its 'piss poor design' and urged, 'Let's just patch the leaky boat.'[5]

The FAA had suffered from successive budget cuts over the previous decade as the Republican party waged war on federal spending and federal agencies. Donald Trump, before the crashes, said he wanted to privatise the agency and scale it back even more.

In a US Senate committee, Senator Richard Blumenthal said Boeing 'decided to do safety on the cheap... and put the fox in charge of the henhouse'. In other words, too much of the safety work on the plane had been outsourced to Boeing itself: 1,000 Boeing employees were seconded to FAA.

---

5 https://www.theguardian.com/business/2020/jan/09/boeing-737-max-internal-messages

What the report reveals is a change in safety culture.

'Boeing and FAA share responsibility for the development and certification of an aircraft that was unsafe,' the report stated:

> For two brand-new airplanes, of a brand-new derivative model, to crash within five months of each other was extraordinary given significant advances in aviation safety over the last two decades. The Max crashes were not the result of a singular failure, technical mistake, or mismanaged event. They were the horrific culmination of a series of faulty technical assumptions by Boeing's engineers, a lack of transparency on the part of Boeing's management, and grossly insufficient oversight by the FAA – the pernicious result of regulatory capture on the part of the FAA with respect to its responsibilities to perform robust oversight of Boeing and to ensure the safety of the flying public. The fact that a compliant airplane suffered from two deadly crashes in less than five months is clear evidence that the current regulatory system is fundamentally flawed and needs to be repaired.

Mary Schiavo, a former inspector general at the US Department of Transportation said, 'At the FAA, they know they're outgunned by Boeing.' She continued, 'They know they don't have the kind of resources they need to do the job they're tasked with doing. They pretend to inspect, and Boeing pretends to be inspected, when in fact Boeing is doing it all almost entirely by itself... So, they got sloppy. Or sloppier.'[6]

Boeing is not alone in having dangerous lobbying power in the US. Amazon, Google and Facebook also lack challengers, constraints on their power and tough independent regulators.

---

6 https://www.bbc.co.uk/news/extra/IFtb42kkNv/boeing-two-deadly-crashes

If only the 737 MAX pilots had been told how to deactivate MCAS: turn the thumb switch on the control column to OFF, set two switches on the centre console to CUT OUT, then adjust the manual stabilizer wheel to trim the aircraft.

And if only Boeing had discarded the legacy of a short metal stairway.

Chapter 20

On the Surface

Jet engines and gas turbines have also been used for surface transport but with mixed results. We have already met the *Lucy Ashton*, the Clyde paddle steamer with her four Rolls-Royce Derwent jet engines.

A Metropolitan-Vickers Beryl turbojet engine normally found in a Gloster Mcteor was squeezed into Donald Campbell's *Bluebird* (K7) jet-powered hydrofoil. This revolutionary craft added 100mph (160km/h) to the water speed record, raising it to over 276mph (444km/h). In his attempt to raise it to over 300mph (480km/h), Campbell was killed when *Bluebird* flipped at high speed on Coniston Water on 4 January 1967.

HMS *Exmouth*, a Type 14 Blackwood-class frigate was used to test the idea of using gas turbines (not jets) for ship propulsion. In 1966, it became the first British warship to be powered by gas turbine engines when it trialled the new Marine Rolls-Royce TM3B Olympus.

The Sheffield-class or Type 42 guided-missile destroyers ordered by the Royal Navy in 1968 used two Rolls-Royce TM3B Olympus and two Rolls-Royce RM1C Tyne marinised gas turbines, driving five-bladed propellers.

These used a clever 'combined gas or gas' (COGOG) propulsion system: a highly efficient but relatively low-powered turbine (the Tyne) is used during economical cruising, while the high-powered Olympus is used during high-speed dashes during combat. The small turbine running at 100 per cent of its power is more fuel-efficient than a large turbine running at 50 per cent of its power. At top speed, the Type 42 could burn 15 tonnes of fuel per hour.

The Royal Navy's Broadsword-class or Type 22 frigate used two Rolls-Royce Marine Spey gas turbines with 37,540shp (28 MW) and two Rolls-Royce Tyne RM3C cruise gas turbines 9,700shp (7.2 MW). The marine Olympus engines became an export success.

In 1986, in an attempt no doubt to reassure the British taxpayer that their pounds were being spent wisely, Lieutenant Commander Nigel Williams water-skied behind Type 22 frigate HMS *Brave*, wearing little more than his uniform cap and a winning grin. Afterwards, he was severely reprimanded by the Flag Officer, Gibraltar. During the stunt, the ship ran its big engines and the crestfallen Williams commented, 'The Rolls-Royce Spey engines have more than enough punch to pull a water-skier out of Gibraltar Bay.'

How about jets for cars? In 1950, Rover used jet technology gleaned from its 1940 contract with Whittle to build the world's first car powered by a gas turbine engine, not a jet. However, it sported the registration number JET 1.

It was unveiled at the Silverstone racing circuit on 9 March 1950. Based on a staid P4 'Auntie' Rover usually bought by doctors, JET 1 looked like a vicar after a heroin bender. Gone were the four doors, replaced by two. Gone were the roof and windscreen, replaced by a low aero screen. Gone were the bumpers. And instead of a plodding piston engine under the bonnet, it mounted a Rover gas turbine behind the two seats, with enormous exhausts. The engine ran at 50,000rpm instead of 4,000rpm, and in 1952, it set the world speed record for a gas-turbine-powered car with a speed of over 150mph. The combination of a gas turbine engine and cart springs in one vehicle must surely be unique.

The Rover designer Spencer 'Spen' King[1] took the car out for an early spin for the benefit of the television cameras. The BBC reported

---

1 The designer of the iconic Range Rover, among other Rover cars.

him as saying, 'I should like to say first of all this car is very easy indeed to drive. I have driven it quite a bit beforehand and there is no difficulty at all. You have two pedals, one to start and one to stop with, and the car handles very nicely. It runs dead straight however fast you're going.'

The car would run on virtually anything: petrol, paraffin or diesel oil, but consumption was heavy at 5 miles per gallon (2km per litre).

'It is obviously the Rover Company's intention to produce a gas turbine engined car as a marketable proposition if and when that becomes practicable,' said chief engineer Maurice Wilks. 'It will certainly be as good as, most probably better than, existing piston engined cars in respect of performance and weight. Probably though, it will not be quite so good in respect of fuel consumption, but to balance that the performance will be superb.'

Rover followed up with two more road cars: the four-wheel drive T3 and the T4, which introduced the Rover 2000 body shape. It also developed a successful business making gas-turbine Aircraft Auxiliary Power Plants (AAPPs), which is a fascinating story in its own right.[2]

Jet cars didn't catch on, though, and there were a number of reasons. Anyone who has driven an early turbocharged car knows that when you press the accelerator... it... takes... ages... to... respond... then SUDDENLY THERE'S TOO MUCH POWER AND NOW YOU'RE IN THE MIDDLE OF A WET CORNER!

This is just the nature of gas turbines. All that rotating mass of blades and turbine discs must gather speed, and then the rate of acceleration itself accelerates, whereas a piston engine is far more throttle responsive and more linear in delivery, too.

All this became evident when a gas-turbine powered Rover-BRM racing car ran at Le Mans in 1965, driven by Jackie Stewart and

---

2 Barnard, Mark, *Pistons to Blades: Small gas turbine developments by the Rover Company* Rolls-Royce, Heritage Trust, Derby (2003). Highly recommended.

Graham Hill, who said, 'You're sitting in this thing that you might call a motor car and the next minute it sounds as if you've got a Boeing 707 just behind you, about to suck you up and devour you like an enormous monster.'

Rover and the British Racing Motors (BRM) Formula 1 team had joined forces to produce the gas-turbine-powered coupe, which entered the 1963 Le Mans 24-hour race as an experimental car, driven by Graham Hill and Richie Ginther. The chief design engineer was Joe Poole:

> Rover had been developing gas turbines for a long time, because they were involved with Whittle with the first gas turbine, but then lost the job and it all went to Rolls-Royce. Rover decided to carry on with turbines, thinking that there was a possibility that they could be used in cars and industrial applications. That's how it all started. The first car that they built with a gas turbine was JET 1, then a number of other cars were produced to test the concept.[3]

The Rover's gas turbine engine was fairly simple, with a single Whittle-style centrifugal compressor, a single combustion chamber and a free-spinning turbine driving the output shaft, separate from the turbine that drove the compressor. The output shaft drove a single-speed trans-axle. The engine was rated at 150bhp. It had a top speed of 142mph (229km/h), but the fuel consumption was 6.97mpg.

'Fuel consumption was pretty dismal,' said Poole, 'so all the time we were trying to develop heat exchangers which would convert the excessive temperature of the exhaust gas and turn that into useful energy, so you needed less fuel. It also idled at about 20,000rpm. The maximum speed was 65,000rpm and you've got to accelerate it between those points. There's not much power to start with, so response wasn't that good.'

---

3 *Autocar* (15 June 2015)

In an attempt to improve the efficiency, and therefore the fuel consumption, Rover fitted ceramic rotary regenerators to the gas turbine. These rotated at only 20rpm. The exhaust gases passed through a section of the honeycomb, which heated up. The disc continued to rotate and then cold inlet air passed through the hot section, heating the air up. This is a clever form of heat exchanger and only slows the gases slightly. There were problems with this new technology, though:

Because of the high temperature difference throughout the heat exchanger, metals weren't the best materials to use. Ceramics were the answer to the whole problem, which is why we went to them. Things didn't quite work out as well as they should have done, and they used to break up quite easily. It was a circular disc with a very fine honeycomb and the air from the compressors would go in one way, picking up heat from the slowly rotating disc, then the hot gas from the turbines went through the other way to heat up the disc. Sealing was another problem, because you'd got high and low pressure across other sides of the disc. But the main problem was the failure of the ceramic material, as it kept breaking up.

During the 1965 Le Mans race, the turbine sucked in track-side sand during the race which damaged some of the turbine blades. The engine began to overheat, and later, while Jackie Stewart was driving, the tip of a turbine blade broke off and damaged one of the regenerators with a big bang. However, the engine kept on running and Stewart kept going to the finish.

Considering the engine damage, the Rover-BRM did well, coming tenth overall and being the first British car home. The average speed was just under 100mph (160km/h) and the fuel consumption had improved to 13.51mpg. This was within sight of a piston-engined performance, at an early stage of development.

But overall, the Rover-BRM was not yet competitive, being slow to accelerate and then difficult to control. At one point, Stewart was overtaken by an 1147cc Triumph Spitfire, also running at Le Mans that year.

'That would have been under acceleration,' explained Stewart. 'Once you got it wound up, it was quite quick. It was pretty good aerodynamically.'

The other problem was the expense. Rover did a cost exercise and found that the turbine materials cost five times that of those for the same power of piston engine. Ceramics were the way to go, but it was too early: a distant echo, perhaps, of Whittle's metallurgical problems in the 1930s.

Politics intervened once again, with Leyland motors taking over Rover. Joe Poole explained:

> We were still working on it when we were taken over by Leyland, who stopped work on the cars because they wanted a bigger engine for truck use. We did the design of a 350bhp unit for them and three or four trucks were built for use by the oil companies, which ran around for a little while, but once again the heat exchanger was the big issue. The real problem was that we were ahead of the available materials that we really wanted. Eventually it was all stopped, and that was the end of it.

It is probably just as well that gas turbines for cars didn't catch on. Fuel economy has become more and more important since 1965, and traffic jams on hot days with the billowing heat from gas turbines would have become intolerable.

Chrysler experimented with turbine-powered cars from 1953 to 1988, producing 50 Chrysler Turbine cars in 1963. Hearing they would run on any fuel, the President of Mexico, Adolfo Mateos, asked if the car would run on tequila? A quick check with the Chrysler laboratory and he was assured it would. A few gallons were sloshed in the tank, and he was given a quick spin. Hairspray or Chanel No. 5 would have done as well, apparently.

How about jet-powered *flying* cars? The Firenze Lanciare has been designed by an ex-US Navy pilot and looks more feasible than the old prop-powered flying cars of the past. The slinky Lanciare can outrun a Ferrari on the tarmac, sprout a pair of wings, light up a pair of jet engines and take to the air at up to 500mph (800km/h). It will have a range of up to 850 miles (1,360km) and land without flaring, on extended shocks. All this will be achieved with the pilot sitting in a luxurious interior, not the usual utilitarian metal box of a fighter jet.

Of all the planned features, the engines are the most achievable – being the Williams FJ33, which is already on the market and used in ultra-light planes. It weighs less than some pilots at 300lb (140kg) and develops between 1,000lbf (4,400N) and 1,800lbf (8,000N) static thrust. Specific fuel consumption at 1,200lbf (5,300N) thrust is 0.486lb/(lbf·h).

Whittle's first W1 engine weighed around twice as much, developed half the power and used about three times as much fuel –1.376 lb/(lbf·h). But he wasn't far off if you compare jet engine development with that of the old piston engines.

Why not attach jet engines to trains? The M-497 Black Beetle is the reason why not. In 1965, Don Wetzel, an engineer for the New York Central Railroad, was given the task of making its trains run faster. His solution was to strap two second-hand Convair B-32 bomber engines onto the roof of an old locomotive. This beast hit 196mph (315km/h) before they could stop it, and a shaken Wetzel decided that the noise and its gargantuan thirst for fuel ruled it out.

Gas turbines for trains was a much better idea, one used by the British Advanced Passenger Train (APT), a tilting high-speed train built by British Rail during the 1970s. At first, Leyland 2S gas turbines were used, derived from the Rover units that raced at Le Mans. The British press lampooned the APT and attacked it for profligacy, despite it only costing £50 million compared with the £100 million British Leyland spent on the Austin Mini Metro. Despite having many features copied by the Class 390 Pendolinos, the APT, like so many other promising British designs, was ridiculed by the press, then cancelled by the politicians.

And the future of jet-engined airliners? H. G. Wells might have speculated about jet fuel made out of hydrogen and carbon dioxide taken from the air, or cold-fusion clean nuclear engines that never need to land.

Then, in September 2020, right in the middle of the Covid-19 pandemic, the FAA signed off the world's most powerful new jet engine.

The GE9X engine is made for Boeing's new widebody 777x aircraft. Each engine produces no less than 105,000lb of thrust for a total of 210,000lb, and during development was tested up to 134,300lb of thrust – a new record. Compare that with Frank Whittle's first jet engine's power of 1,000lb. The whole engine is 15ft (4.6m) in diameter and the fan has a diameter of 134in, or 11ft (3.4m) and has 16 curvy carbon-fibre blades. The RB-211 design was finally justified.

The turbines reach a temperature of 2,400°F, the temperature of lava, and use ceramic matrix composite materials, which, after decades of struggle, are at last proving reliable enough.

And in May 2021, Rolls-Royce opened a testbed for its new Ultrafan jet engine, which will have an even bigger fan diameter than the GE9X at 140in (3.55m). This monster will suck in the equivalent of an Olympic-sized swimming pool of air every second and will be 25 per cent more fuel-efficient than the first-generation Rolls-Royce Trent engine.

The amount of fuel that is added to the compressed air in today's jet engines is still short of what it could be for complete combustion of the oxygen in the air supplied. So there is still more power to be had – if the world of jet travel still wants it.

But political pressure to end the burning of fossil fuels is rising across the world. The Rolls-Royce Ultrafan jet engine might well be the last bellow of a dinosaur.

Shrinking the World

Imagine, if you will, a modern nation on the brink of war with a neighbour, a war your leaders know will be won or lost in the air. One of your serving pilots invents a new powerplant that will quadruple the speed of your aircraft and, after the war is won, will give you a ten-year lead in civil aviation. Wouldn't you support such an individual?

Well, no. Not only did his superiors ignore, denigrate and downplay Whittle's new jet engine, after building an aircraft that proved the idea, they then gave away his technology not only to their American allies but also to their Cold War enemies in the Soviet Union.

Surely never before in the history of technology was so much given away by those who understood so little. This theme continued, with Britain's Miles M-52 supersonic project cancelled after vast amounts of taxpayer's money was spent, then the TRS-2, then the gas-turbine-powered APT.

This still goes on: Britain's politicians and decision makers continue to make the wrong decisions about large projects. C. P. Snow believed that the corrosive effect of Britain's old caste system had held back our finest scientists and engineers: Victorian schooling had overemphasised the humanities in the form of Latin and Greek at the expense of scientific education, and this bias seems to continue. Britain's politicians still don't seem to know much about science or engineering. Neither do they seem to know much about finance. Confucius (551–479 BCE) knew the importance of educating a nation's officials, but Britain has still to learn this lesson.

We have seen how the jet engine took over from the piston engine, and how the numbers of passengers rocketed after jet airliners became acceptable. The roots of the change, as with so much in today's Western culture, lay in the 1950s, and 1957 can be seen as the tipping point in the US when the numbers travelling by air across the country exceeded those travelling by rail for the first time.

Britain, with its huge post-war aviation industry briefly held sway with the revolutionary de Havilland Comet, crossing the Atlantic in 1958. Then the bigger and cheaper Boeing 707 overwhelmed the Comet. The 707 had a price double that of the piston-engine aircraft it would replace, but it would produce three times the revenue. The American industry's economies of scale beat the British aviation industry to its knees.

The transatlantic liners saw the writing on the wall and attempted to reduce the time spent at sea between the Old and New Worlds, the record falling to three and a half days by 1952. But, by 1960, the numbers crossing the Atlantic by jet plane exceeded those travelling by sea and the Age of the Liners was over.

The advent of the Boeing 747 and other wide-bodied airliners powered by high bypass turbofan jets democratised long-distance air travel. The price of tickets fell to the point where ordinary First World people could hope to fly abroad for their holidays.

One consequence of the coronavirus pandemic is that the old national airlines might get their comfortable high-fare world back.

As a result of the low fares, the seaside towns of Britain, like the ocean liners before them, and the railway hotels before them, fell into disuse and disrepair. And all because of the jet engine.

Humanity marched on as ever, discarding the old things behind.

Bibliography

Barnard, Mark, *Pistons to Blades, Small gas turbine developments by the Rover Company*, Rolls-Royce Heritage Trust, Derby (2003)

Baxter, Alan, *Olympus*, Rolls-Royce Heritage Trust, Derby (2007)

Blackman, Tony and Wright, Anthony, *V Force Boys: All New Reminiscences by Air and Ground Crews Operating the Vulcan, Victor and Valiant in the Cold War and Beyond*, Grub Street, London (2017)

Boyne, Walter, *The Jet Age: Forty Years of Jet Aviation*, National Air and Space Museum, Washington, DC (1979)

Bryen, Stephen D., *Technology Security and National Power: Winners and Losers*, Routledge, Abingdon (2017)

Budiansky, Stephen, *Code Warriors*, Alfred A. Knopf, New York (2016)

Campbell-Smith, Duncan, *Jet Man: The Making and Breaking of Frank Whittle, Genius of the Jet Revolution*, Head of Zeus, London (2020)

Calder, Simon, *No Frills*, Virgin, London (2006)

Conner, Margaret, *Hans von Ohain: Elegance in Flight*, American Institute for Aeronautics and Astronautics, Inc., Michigan (2001)

Constant, Edward, *The Origins of the Turbojet Revolution*, The Johns Hopkins University Press, Baltimore (1980)

Golley, John, *Genesis of the Jet*, Airlife, Marlborough, Wiltshire (1996)

Gunston, Bill, *Faster Than Sound*, Patrick Stephens, Wellingborough (1992)

Hamilton-Paterson, *Empire of the Clouds*, Faber, London (2010)

Heathcote, Roy, *The Rolls-Royce Dart*, Rolls-Royce Heritage Trust, Derby (1992)

Haworth, Lionel, *The Rolls-Royce Tyne*, Rolls-Royce Heritage Trust, Derby (2000)

Howard, Bill, *The Jet Engine Story*, Farnborough Air Sciences Trust, Farnborough (2019)

Hoyland, Graham, *Merlin: The Power Behind the Spitfire, Mosquito and Lancaster*, William Collins, London (2020)

Hughes, Ted, 'Work and Play' (1961)

Kaplan, Philip, *Big Wings: The Largest Aeroplanes Ever Built*, Pen and Sword, Barnsley (2005)

Knowles, Arthur, *The Bluebird Years*, Sigma, US (2005)

Owen, Kenneth, *Concorde: Story of a Supersonic Pioneer*, London: Science Museum, London (2001)

Ruffles, Philip, *The History of the Rolls-Royce RB211 Turbofan Engine*, Rolls-Royce Heritage Trust Derby (2014)

Shute, Nevil, *No Highway*, William Heinemann, London (1948)

Stokes, Peter, *From Gypsy to Gem*, Rolls-Royce Heritage Trust, Derby (1987)

Taylor, Douglas R., *Boxkite to Jet*, Rolls-Royce Heritage Trust, Derby (1999)

Whittle, Frank, *Jet*, Muller, London (1953)

Wells, H. G., *The War in the Air* Macmillan, London (1908)

Wells, H. G., *The World Set Free*, Macmillan, London (1914)

Yeats, W. B., *The Second Coming*, The Dial, Dublin (1920)